# Anticipation Science

Volume 7

**Anticipation Science** encompasses natural, formal, and social systems that intentionally or unintentionally use ideas of a future to act in the present, with a broad focus on humans, institutions, and human-designed systems. Our aim is to enhance the repertoire of resources for developing ideas of the future, and for expanding and deepening the ability to use the future. Some questions that the Series intends to address are the following: When does anticipation occur in behavior and life? Which types of anticipation can be distinguished? Which properties of our environment change the pertinence of different types of anticipation? Which structures and processes are necessary for anticipatory action? Which is the behavioral impact of anticipation? How can anticipation be modeled?

The series is interested in receiving book proposals that:

- are aimed at an academic audience of graduate level and up
- combine applied and/or theoretical and/or philosophical

studies with work especially from disciplines within the human and social sciences broadly conceived. The series editors aim to make a first decision within 2 months of submission. In case of a positive first decision the work will be provisionally contracted: the final decision about publication will depend upon the result of the anonymous peer review of the complete manuscript. The series editors aim to have the work peer-reviewed within 4 months after submission of the complete manuscript. The series editors discourage the submission of manuscripts that are below 150 printed pages (75,000 words). For inquiries and submission of proposals prospective authors can contact the editor-in-chief:

Roberto Poli: roberto.poli@unitn.it

Jim Dator

# Beyond Identities: Human Becomings in Weirding Worlds

Springer

Jim Dator
Political Science
University of Hawaii at Manoa
Honolulu, HI, USA

ISSN 2522-039X ISSN 2522-0403 (electronic)
Anticipation Science
ISBN 978-3-031-11731-2 ISBN 978-3-031-11732-9 (eBook)
https://doi.org/10.1007/978-3-031-11732-9



This Springer imprint is published by the registered company Springer Nature Switzerland AG
The registered company address is: Gewerbestrasse 11, 6330 Cham, Switzerland

# Preface

This book has one theme—identity—but many, many parts, from many disciplines and perspectives, from the past and present to very distant futures. It may annoy specialists and activists who find I fail adequately to express the main features of their field. I may offend people who think I should not express my own views so sharply. But I hope it will encourage the many young people who inspired me to continue becoming, and not to be thwarted by those who insisted they should be what they are not.

Please read this book as you would read the score of a complex symphony, or as you would listen to such a symphony being played from the first, fresh note, on and on through many moods, motifs, expositions, repetitions, inventions, tempi, timbres, cadences, crescendos, diminuendos, and movements.

Consider especially the movements. The classical symphony has four movements, though there may be more, or less, each with its own characteristics both independent of and interdependent with the other movements. Each movement could stand alone as a harmonious whole requiring no further elaboration. Sometimes, a common theme or its variations is expressed across all movements. Sometimes, each movement appears quite separate from the others, and yet in the end, when the last note is played, when it exists now entirely in your memory, if at all, the movements form an interwoven whole. It is a fully integrated piece of work from many strands.

Specifically, you might consider the structure and motif of this book to be a symphony by Mahler or Bruckner—long, loud, sometimes droning on and on in prolonged, sustained chords, barely moving, just blasting away into the brainpans of the listener, or else so soft, so strained, the listener must also strain to perceive anything at all, much less to comprehend it. What are those long, sustained passages for? What relation do they have to the main point of the main theme? And what about the poems—the songs? At the least, they prolong a recapitulated mood, a feeling, heightening variations on the themes. Accept them on their own terms. Do not juxtapose them against each other. They will magnify the sense of the whole when the movement, the symphony, finally comes to its end.

*Personal History*
*by Kareem Tayyar*

This was in the year when
a ship took leave of the water
& floated out across the clouds.

When the clouds became
the opened palms of the angels.

The year when the angels strung
their wings across the telephone lines
like laundry drying in the sun.

Only there was no sun.
Not that year.

That came later,
when the children turned first
into horses & then into ghosts,

when the rain fell in love
with a poet,

when the poet forgot his own name
& then the names for everything else.

That was a good year.
A year without names.

A year when I learned to kneel
without my knees ever touching the ground,

& where the gods all prayed to us.

But I am getting ahead of myself.
I should start at the beginning.

*Poetry*, Vol. 219, No. 4, January 2022, p. 323
Used with permission from the author.

Honolulu, USA Jim Dator

# Contents

# Chapter 1
# What I Am Not, and Why

**Abstract** Reared by three women, in the US South, of uncertain paternal ancestry, rejecting identities ascribed to me against my will—White, male, Southern in particular—while not adopting other identities instead, I explain why I am not a human being but rather a human becoming constantly creating and re-creating myself on the basis of experiences in various places in the world, especially Japan and Hawaii.

**Keywords** Becoming · Being · Human · Identity · Japan · South · Yugoslavia · Verb · World futures studies federation

Appiah (2018: iv) writes, "My main message about the five forms of identity [creed, country, color, class, culture] is in effect that we are living with the legacies of ways of thinking that took their modern shape in the nineteenth century, and that it is high time to subject them to the best thinking of the twenty-first." I by no means place myself among the best minds of any century, but what I have written here is an argument for moving beyond identity. Identity as it is most often currently used as the necessary foundation of a true, authentic, non-alienated, essential, culturally/historically/ethnically/biologically-grounded self is pathologically destructive. It is the cause of much individual and collective misery, anxiety, grief, and violence today and for future generations. That concept of one's identity derived from the past needs to be rejected. One's identity can and must be derived from the futures, not as a human being but as a human becoming.

Instead of having an identity forced on us at birth, or of spending time and energy trying to discover our identity in the past and/or in some group in the present, we should and can create values, processes and institutions that enable each of us, from childhood onward, continuously to create identities that are pleasing to ourselves and others, dynamic, interactive, and futures-oriented; that instead of striving to be human beings, we should create ourselves as continuously metamorphosizing human becomings.

Moreover, our identities are typically framed in terms of opposition to other identities that we have been taught to disparage and hate on the basis of past grievances that continue to fester in part because we actively pick at the scabs.

David Bromwhich observes that "(i)t seems to me that if you're taught that you are the product of a culture defined by nation, race, region, religion—that you are



J. Dator, *Beyond Identities: Human Becomings in Weirding Worlds*,
Anticipation Science 7, https://doi.org/10.1007/978-3-031-11732-9_1

rooted in that culture and the culture is what constitutes you—then it becomes a kind of treachery, a kind of treason against who you are, to want to pull yourself away from the culture you belong to by birth. Whereas tradition can be a matter of change and of innovation. Tradition gets bent to new purposes by people who are versed in what the wisdom of their predecessors may add up to, but who also, for reasons moral, political, or aesthetic, want to do something different" (Gutkin 2022).

Recently, many discussions of identity have centered on people who seek to resist and discard their assigned identity and adopt a new one. This is very impressive. It shows I am not alone in this quest. I offer detailed examples of such identity-fluid seekers. However even they seem obliged to choose a new identity among an array of existing ones. I have seen only a few, exciting examples of people who are trying entirely new identities of their own invention. Everyone should be encouraged and enabled to experiment continually throughout their lifetimes. A feature of life from childhood on might become a series of trials and experiments with different identities, some with long histories while others are tentative novel self-inventions.

Appiah (2018: 217) writes, "There is a liberal fantasy in which identities are merely chosen, so we are all free to be what we choose to be. But identities without demands would be useless to us. Identities work only because, once they get their grip on us, they command us, speaking to us as an inner voice; and because others, seeing who they think we are, call on us, too. If you do not care for the shapes your identities have taken, you cannot simply refuse them; they are not yours alone. You have to work with others inside and outside the labeled group in order to reframe them so they fit you better; and you can do that collective work only if you recognize that the results must serve others as well". I agree with much of what he says here, but disagree that we have no choice but, in cooperation with others, to loosen the bonds of our assigned identities so they are tolerably less restrictive and proscriptive.

I believe that fluid, futures-oriented identities orientation are possible, desirable, healthy, and exhibited in the lives of a growing number of people who have rejected their assigned identities and assumed new ones. Moreover I will draw to some extent from my own experience as a human becoming. I was fortunate to have gained consciousness as a fatherless child of uncertain ethnicity reared by three strong, capable women, supported over time by a vast array of people in various socializing institutions—religious, educational, athletic, multicultural.

Though I was born in 1933 in rural, upper New York State when my mother attended Eastman School of Music in Rochester, I was powerfully and deeply influenced by having spent my childhood and early adulthood in the racist, xenophobic, patriarchal, evangelical/fundamentalistic, violence-loving, unreconstructed South (panhandle and central Florida, Appalachian Georgia, North Carolina, and Virginia) from 1933 to 1959 when I finally left those parts of the world for good. To be sure, I was also powerfully inspired by southerners I knew (especially several Anglican priests and Stetson University professors) who were not racist, xenophobic, patriarchal, fundamentalistic, or violence-loving. I was impressed most powerfully of all as a youth and young adult by people who were the victims of racism, xenophobia, patriarchy, fundamentalism and violence that I did not personally experience. I was an out-spoken nigger lover from the earliest times.

I absolutely did not—and do not—identify with, or want to be identified with, the White, male, culture of the US South. Nor did they want me to be a part of them either! Without desiring or pretending I was something else, I did my best not to be White, male and Southern—because in fact I wasn't objectively clearly any of them. It was others who said I was male, White, Southern. That was not my identity, chosen or intentionally displayed. Indeed, I prided myself in having no familial identity whose destiny I must fulfill. I was a human becoming.

I was significantly reinforced and re-formed during my first fulltime job with a fresh PhD—teaching political science at Rikkyo University in Tokyo, Japan, during the early 1960s. Japan is a culture into which it is almost impossible for non-Japanese to assimilate. No matter how I acted—whether doing what was most normal or most outrageous to me—I was equally weird to most Japanese. No matter how well I spoke Japanese (never very well), or tried to behave as other Japanese behaved, I was always *hen na gaijin*—strange foreigner. At the same time, Japan was an open, safe, futures-seeking society while I was there. My family and I were never threatened or discriminated against. My three very young blond and red-headed children freely roamed all over Tokyo on the subway, rail, and bus transit system in complete safety, though riders would sometimes touch the hair on their head and arms in wonder.

Living in a culture strongly influenced by Confucian concepts, we had our place, with certain rules, duties, responsibilities and expectations that we eventually learned and endeavored to fulfill. At the time, and I think still, most Japanese ranked themselves against foreigners—foreigners were not equal in weirdness. Some foreigners were rather close to Japanese. Others stood very far away (Hayashi 1961–82). As a reputedly White, male, American professor in a prestigious private Japanese university, I ranked relatively high. In some ways too high: It was as though I was perched precariously one notch above and to the side of ordinary humans; I was a kind of angel. Nice to have around to grace certain formal or auspicious occasions, but, like any angel, a priggish annoyance when humans want to be human. Then, I needed to bow politely and take my leave. So in fact I did experience serious discrimination, but that of a kind of bland, superficial superiority rather than of the wounds and scars of a history of inferiority, enslavement, murder, and abuse.

My birthday is August 15—the day Japan surrendered at the end of World War II. "Should I dare celebrate my birthday," I asked my Japanese friends. "Of course!" they replied. "August 15 is the best day in Japanese history! It should always be celebrated!" And indeed, when serious consideration was given in the 1960s for Japan to choose a day to celebrate its nationhood, August 15 ranked high among the dates considered, though it was not chosen. Having lived most of my life to that point in the US South where the War of Northern Aggression was still being actively fought, such acceptance of defeat and a deep desire for new directions was beyond belief. "We Japanese were profoundly wrong, and needed severely to be corrected," my friends said. "The way to be successful in the world is not through military domination! What were we thinking? That was so old-fashioned. The way to rule is through economic domination," and so they set out to make *Japan as Number One* (Vogel 1979).

Nonetheless, the identity Japanese assigned to me made no sense to me whatsoever. I did not want to be categorized in any way, especially in a way that was to my undeserved advantage sometimes. I yearned to live in Japan with all the privileges and obligations—oh, the obligations!—of an ordinary Japanese, but the Japanese understood that was impossible: I would not assume those obligations, and the Japanese knew it from long experience with foreigners. So I played my angel role as best I could and, sure enough, when Sirens sang from distant shores, we left.

I learned many things from my experiences in Japan. One was to ignore what others thought of me without my being purposely rude or irresponsible. Eventually, I came to realize that almost never did anyone think anything of me at all! Each of us is intently thinking about ourselves, and about what others think of us, which, of course, is nothing, or seldom anything of consequence. So while I did not set out to insult Japanese, or Americans, or anyone, I no longer cared what people thought of me, because I knew they weren't thinking of me, and when they did, I could probably find a way to carry on or acquiesce: no point in my being an impertinent jerk. Smile, nod, turn away. This awareness that I gained first in Japan remained with me and strengthened for the rest of my life.

Nonetheless, I did what I could to minimize what advantage being White male might give me not only in Japan but throughout my life. I couldn't easily become Black or female, but I could become an offensively unsightly White male. My skinny ugly hair was always a problem that anyone who wanted to bug me could complain about. After one criticism too many, I got a crew cut in high school, causing endless negative comments. Indeed, in Japan, barbers wondered what defeat I had suffered since athletes losing a contest frequently shave their heads—or have them shaved. So I let my hair grow as long as the scrawny threads could dangle which caused far more controversy, first as a dirty hippie. Then, as I porked up with age, I was frequently addressed as "ma'am" by people who assumed I was a fat old woman—an insult to fat old women worldwide, perhaps, but not to me. Almost the first question I get everywhere in the world after a lecture is why I wear my hair as I do—unchanged (save for color, from brown to gray) after 50 years. I refused to wear a coat and tie no matter how august the occasion, preferring clean but disreputable clothing. My motto is "Always make a bad first impression", and short of staggering on stage drunk with my fly open, I usually do. My point being, judge me by what I say and do, and not how I look.

I felt vindicated by an episode of *South Park* mimicking *Wheel of Fortune* that asked a contestant to fill in the missing letters to answer the question, "what do you call people who annoy you?" The contestant balked when he got all letters but one: N? GGERS insisting he couldn't say the word on television. He was repeatedly assured it was perfectly all right for him to say it, which he did—losing in disgrace because the correct letter was A and people who annoy you are NAGGERS.

I was also inspired to move beyond being merely human by a few specific individuals, especially Buckminster Fuller who proclaimed "I am not a human being; I am a human becoming". "I live on Earth at present, and I don't know what I am. I know that I am not a category. I am not a thing—a noun. I seem to be a verb, an evolutionary process—an integral function of the universe" (Fuller et al. 1970). This struck me powerfully, perhaps in part because while nouns and pronouns—especially "I"—figure prominently in the English language, Japanese is mainly verbs, adverbs and adjectives. In Japanese, one has a strong sense of what is going on, and how to feel about it, but often very little clarity as to whom is acting on whom. And so maybe I am a human becoming, more a gerund than a noun.

The economist, W. Brian Arthur, also came to realize that economics typically "expressed via algebraic mathematics, deals only in nouns—quantifiable nouns—and does not deal with verbs (actions), and that this has deep consequences for what we see in the economy and how we theorize about it" (Arthur 2021). Biology, he discovered, points to a better way. "Biology is a procedural discipline, not based on quantities growing and changing, but rather on processes: processes that determine the step-by-step formation of structures…; processes that respond to their internal and external environment…; processes that create novelty…." It may be that in essence all the social sciences are like economics—mainly nouns with few verbs. This seems to include many of the people in the various disciplines concerned with identity. The exceptions seem to be those who are process-oriented and understand identity as dynamic "processes that create novelty".

"The natural language of life is algorithmic," Arthur concludes. So, maybe Fuller would revise his sense of self. Following Arthur, Fuller might now say, to emphasize his endless becoming, "I think I am an algorithm."

Fifty years living and teaching in Hawaii has done more than a little to shape my understanding of the attraction and pathologies of identity as well. So did my experiences as Secretary General and President of the World Futures Studies Federation for two decades and then on the faculty of the International Space University for two more decades, both of which took me to about 75 countries, many communist (including North Korea), from the 1970s onward, where I had the extraordinary opportunity of talking with people, high and low, about their hopes and fears for the futures. The 1980s were a decade of turmoil for people who sought to change their identity from communist to social democrat, royalist, chauvinist, or capitalist and I got to see it first-hand almost everywhere.

I obviously cannot make a fair or convincing argument for futures-oriented identities on the basis of my experiences alone! Over my lifetime I have observed that identity—having it, losing it, misappropriating it, recovering it, the centrality of it—has become a major source of unity and harmony as well as division and bloody conflict not only in the Disuniting States of America but in many other parts of the world as well. For a decade, students from the futures program of the University of Hawaii at Manoa and myself went to Dubrovnik, Yugoslavia, every April to participate in a two-week futures course at the InterUniversity Centre for Postgraduate Studies there. It gave us an unequaled opportunity to engage with people from Africa, western and eastern Europe, and the Soviet Union that we could not meet otherwise—people we came to admire and love. We were devastated when "Yugoslavia"—a hopeful, progressive, harmonious bridge between "East and West" (meaning communism and capitalism)—collapsed into ethnic slaughter, including in the 1991–92 bombing by Serbs of the ancient medieval but vibrant city of Dubrovnik (in Croatia) and the destruction of the buildings of the InterUniversity Centre. Absolutely nothing we had experienced in all our times in that magical city led us to understand that the façade of unity of the historically-conflicting ethnicities as a single, peaceful country, Yugoslavia, had been almost entirely held together by the glue of communism. When the Wall fell, old animosities leapt as all-consuming flames devouring years of hope and love that I had witnessed as genuine and permanent. Some of my best friends—gentle, humane people of immense intelligence and wit—instantly became irreconcilable enemies, inhumane monsters or victims in the killing fields.

Yugoslavia and surrounding areas were among the most extreme case of what happened in all formerly communist countries. During the 1980s I met not only ordinary people but presidents of nations and heads of organizations with names like The Marx-Lenin Institute who said their fondest hope was to go shopping at Macy's and eat bananas in winter, which they saw people doing in the "West" on CNN. They spoke to me openly about this. In May 1989 I gave a talk on "Everyday Life in the 21st Century" (Dator 1990) to members of Bulgarian President Todor Zhivkov's Council of Advisors one day while on the next, a lecture on "Futures of Democracy" in a huge hall at Sophia University in Bulgaria where students filled every seat, every spot on the floor and stage around me, as well the open windows. A few days after I left, the students and others took to the streets with increasing frequency, Zhivkov was replaced by someone even more moderate, and the communist government eventually collapsed peacefully. Something somewhat similar happened during a trip to Riga, Latvia, a few months later: we arrived one evening ready to meet with various ministers of state the next day, but the system fell overnight so all our appointments vanished. Somehow, life went on as usual: people who could went to work or shopping; they stopped at traffic lights. There was absolutely no looting or rioting. Amazing, I thought.

Nonetheless, no one who most fervently wanted system change was prepared for the sudden, peaceful collapse of communism when it unexpectedly happened (almost) everywhere. No one was ready and able to take over governance with a vision and action plan in hand for the better tomorrow they dreamed about, fully endorsed by the people. So when the collapse happened there was a vacuum everywhere that allowed local and international agents of capitalist governments and corporations to move in to ensure that the dreams of a more liberal, open, and equitable society had no chance to emerge. Folks might be able to eat bananas from central America in midwinter, and vote, but that was about it. While feelings differ across eastern Europe, there are still significant numbers of people—a majority in some countries—who feel their lives have become worse after the fall of communism, not better (Pew Research 2019). Love—and fear—of Russia never vanished, however, and with the Russian invasion of Ukraine we see some of the fruit of those long-dormant seeds.

This is the futurists' curse: "may your dreams come true!" If you have not asked and answered the question, "what's next?" the chances are pretty high that your dreams of a new identity will be dashed and soon become nightmares. Sometimes old identities revive their appeal long after it is possible to recapture them. You get the worst of the past instead of the best.

And yet at the same time I happily note that in spite of oppressive official measures against them, impressive numbers of people everywhere are questioning identities now, especially who gets to tell them what their identity is, along with what they must believe and how they must behave accordingly. They insist that they can and should chose and proclaim their identity for themselves—sometimes even declaring identity with groups that refuse to accept them. And yet even they seem to assume that they should have some kind of a fixed identity and it is that notion also that I wish to problematize.

## References


Appiah, Kwame Anthony. 2018. *The lies that bind: Rethinking identity*. New York: Liveright Publishing.

Arthur, W. Brian. 2021. Economics in nouns and verbs, 5, April 20. https://arxiv.org/abs/2104.018 682021.

Dator, Jim. 1990. Everyday life in the 21st century. *Sociological Problems*, 2. [In Bulgarian. Published in Sofia, Bulgaria].

Fuller, R. Buckminster, Jerome Agel, and Quentin Fiore. 1970. *I seem to be a verb*. New York: Bantam.

Gutkin, Len. 2022. 'An elaborate new decorum has crept in' David Bromwich on politics, manners, and the therapeutics of identity. *The Chronicle of Higher Education*. https://www.chronicle.com/article/an-elaborate-new-decorum-has-crept-inMarch15.

Hayashi, Chikio. 1961–82. *Nihonjin no Kokuminsei*. Tōkei Sūri Kenkyūjo, vol. 4. Tokyo: Kokuminsei Chōsa Iinkai.

Vogel, Eric. 1979. *Japan as number one: Lessons for America*. Cambridge: Harvard University Press.

Wike, Richard, Jacob Poushter, Laura Silver, Kat Devlin, Janell Fetterolf, Alexandra Castillo, and Christine Huang. 2019. Political and economic changes since the fall of communism. https://www.pewresearch.org/global/2019/10/14/political-and-economic-changes-since-the-fall-of-communism/.

# Chapter 2
# Identity, More or Less

**Abstract** The history of "identity", a relatively new concept that was "lost" almost as soon as it was "discovered", is reviewed. The dangerously pathological features of historical, ascribed identity, and the fact that identities fail to identity, are emphasized. Fluid, futures-oriented identities are preferred.

**Keywords** Becoming · Being · Future shock · Futures · *Homosapiens* · Human · Identity · Immigrants · Nation · *Ningen* · Postmodern

The idea of "identity" is both ancient and new. Countless studies state that identity answers the question, "Who am I?"—as though each of us has a predetermined identity that pre-exists in the past as well as in certain groups and categories of the present. It seems to me that the far more important and urgent question for each of us that we should continuously ask and answer is not "who am I" but, "who do I want to be…, next? How might I continue to become?" Life as a futures quest should be instilled in children from the earliest time, and reinforced by formal ceremonies along one's lifespan, like—but radically different in focus from—baptism, confirmation, bar/bat mitzvah, and other coming of age and preparing to die ceremonies that once may have been sufficient.

I am not saying that people should be taught affirmatively to renounce or disparage their past, much less the past of their parents, caregivers, community, nation, culture, language, religion, and the rest—although all of us do need to be sure we don't live in the romanticized shadow of unexamined realities. It is rather a question of priority, focus, and balance. What is of vital importance is not what we were but what we can or should try to become; not primarily who our ancestors were and what they did (or was done to them), but rather whether our descendants are able to choose how to live their lives as a consequence of how we and others live our daily lives now.

It might not be too much to say that we can "know" the futures better than we can know the past because we have at least some individual and collective agency over the futures, but none over the actual past. We can and do constantly rewrite or reinterpret some few fragments from the past, but whatever "really happened" is beyond our ability either to know in detail or to change at all. We have very little real control over the present as well, given how swiftly the present becomes the future before we can act on it. If we are not encouraged and enabled to be routinely but

 
J. Dator, *Beyond Identities: Human Becomings in Weirding Worlds*,
Anticipation Science 7, https://doi.org/10.1007/978-3-031-11732-9_2

deeply foresightful (and we are not), we become at best reactive, and more likely perpetually bushwhacked.

Far more important it is to understand and act on the understanding of our agency in becoming a self of our choosing by interacting with, learning from, and inspiring other futures-oriented human becomings in co-creating and recreating preferred futures.

The way I have phrased things so far may seem to imply that identity creation is an act of supreme individualism: it is MY identity that I create all by and for myself. If I have given that impression, then let me correct it immediately. Identity should be futures-oriented, dynamic, invented, constructed by each of us interacting with other questing selves, in part on the basis of precedence but mainly on the basis of consequence. This most certainly is not a solitary individualistic task. Quite to the contrary. This is an intensive social activity. The word that can be translated as "human", 人間, (*ningen*, in Japanese), is composed of two characters, 人, which means person, and 間, which means "between", "among", "within", "inside", "space"—that is to say, a human is not a solitary selfish individual but a person among other persons, living, dead, and yet to be born. It is in part by interacting with others, past, present and futures, that we each are human becomings. In his article, "Bricolaging a public philosophy for future generations", Tae-Chang Kim quotes the 20th Century Japanese philosopher, Tetsuro Watsuji, on the meaning of *ningen*: "It is a word that signifies the betweenness of human beings, that is, the public." "According to it, *ningen* is the public and, at the same time, the individual human being living within it" (Kim in Kim and Dator 1999: 261).

*Poem Toward People*
By Ariel Yelen

I've always been obsessed with people—
whether or not I know them. Obsessed
by our knowledge of each other, the quality

of connection, our friendship or non-friendship,
its relation to other connections. Obsessed
by the way a new connection can change pre-existing

ones, reorder them, renew them, fine-tune
or disappear them. By the light pressure
of an other's existence, which in turn grows

me. Obsessed by memory and lack of memory
for the way things were—I don't think I'd recognize
you if I saw you on the street, though in the past

so obsessed I thought almost everyone
was you. Obsessed with leaving people
so I can obsess about them again.

By thinking with and through people, dead
and alive, without whom I'd be a different person,
think different thoughts. Even obsessed

with the version of me I don't know, walking around
having met different people, thinking different
thoughts, moving in a different direction, away

from people and toward the self,
or the desert, or the sea, or the god, or the page, or the mountain,
or the canyon, or the forest, or the dark (Yelen 2022).

Used with permission from the author.

Yes, that is the point: We are who we are, and are not who we are not yet who we might become, because of who we happen to have known or known about as well as all those we have never met and who, if we had (and if we do), would have helped us be a different person—for better or worse. We owe our thanks, respect, and love to all, and not just our biological family, tribe, nation, or species.

That is only the beginning of the journey. We must also consider the possibility and desirability of humans becoming transhumans and posthumans, or rather of the evolution of posthumans from current *homo sapiens, sapiens.* Neither our ancestors nor ourselves now are the ultimate "crown of creation". We are merely one phase in an infinite process of evolutionary metamorphosis, but we can and should strive to be active, caring co-creators within that cosmic process when we have that opportunity.

## 2.1 Identities Lost

We don't know who discovered water, Marshall McLuhan used to say, but we're pretty sure it wasn't a fish—though it might well have been discovered by a fish out of water. We don't know who discovered identity, but few people gave it a thought until they lost theirs and struggled to recapture it—or to create a new one. As Kobena Mercer stated, "identity only becomes an issue when it is in crisis, when something assumed to be fixed, coherent and stable is displaced by the experience of doubt and uncertainty" (Gove and Watt 2004).

In some ways, humans are profoundly migratory—Out of Africa, perhaps, across the globe, sometimes pushing into areas where humans had never been before, but often contesting with and sometimes eliminating other *homo sapiens* who were there before. Nonetheless most people lived and died within a few miles of where their family's and friends' families and friends had lived and died. It wasn't until people were forced by theft, (mis)fortune, or famine to move from ancestral lands and ancestral ways of living that groups and individuals lost their identity. Many struggled in newfound wildernesses often either alone or in fragile nuclear families in competition and cooperation with thousands of strangers who were similarly struggling. This became a fact of life for millions of people when the 1000-yearlong Reich of western Christendom disintegrated when the Renaissance, Black Death, Reformation, and unusually stable, temperate climate enabled new people with new ideas and technologies to create ever-newer worlds. While an argument can be made that the modern sense of "identity" and the "self" arose when Descartes declared *Cogito ergo sum* in 1637, the full weight of identity—or its lack—seems to lie in the 19th Century when globalizing processes, technologies, and ideologies severed people from their hidden selves and required them to create new ones—over and over as the pace of social and environmental change quickened and new strangers showed up to replace old strangers who moved—or were removed—on.

Stuart Hall says that the concept of "identity" developed in three phases. The first was as a subject during the Enlightenment period. The second was as a sociological subject during modernity. And the third is the subject during post-modernity. "The Enlightenment subject was based on a conception of the human person as a fully-centered, unified individual, endowed with the capacities of reason, consciousness and action, whose 'centre' consisted of an inner core which first emerged when the subject was born, and unfolded with it, while remaining essentially the same—continuous or 'identical' with itself—throughout the individual's existence." "The notion of the sociological subject reflected the growing complexity of the modern world and the awareness that this inner core of the subject was not autonomous and self-sufficient, but was formed in relation to 'significant others', who mediated to the subject the values, meanings and symbols—the culture—of the worlds he/she inhabited" (Hall 1992: 276).

The post-modern subject has "no fixed, essential or permanent identity." "Within us are contradictory identities pulling in different directions, so that our identifications are continuously being shifted about. If we feel we have a unified identity from birth to death, it is only because we construct a comforting story or 'narrative of the self' about ourselves" (Hall 1992: 277).

Similarly, Philip Gleason says that the authoritative *Oxford English Dictionary* (OED) defined "identity" as "the sameness of a person or thing at all times or in all circumstances; the condition or fact that a person or thing is itself and not something else; individuality, personality." However, "The original *Encyclopedia of the Social Sciences*, published in the early 1930s, carries no entry at all for identity", while "*The International Encyclopedia of the Social Sciences*, published in 1968, has a substantial article on 'Identity, Psychosocial,' and another on 'Identification, Political'" (Gleason 1983: 910–931).

Beginning in the 1950s in the US, works decrying and searching for lost identity were in flood tide by the 1960s: David Riesman, Nathan Glazer and Reuel Denney *The Lonely Crowd* (1950)—a book that made a great impact on me, especially the distinction between inner- and other-directedness; Will Herberg, *Protestant-Catholic-Jew* (1955); C. Vann Woodward, "The Search for Southern Identity," (1958); Peter Berger and Thomas Luckmann, *The Social Construction of Reality* (1966)—another very impactful book for me; William Glasser, *The Identity Society* (1972)—which contributed to my fascination with human evolution—and many others. One of the most influential writers overall, Eric Erickson, coined the phrase, "identity crisis" in his book, *Identity: Youth and Crisis* (1968).

The 1950s–70s was also a period when specific groups who felt categorically and historically excluded from mainstream American identity began to reorganize for integration and equality, or for recognition and freedom, most notably women and Negroes, as they were sometimes called then. Identity angst now is global, often specifically inspired by and contributing to local versions while spawning novel identities.

## 2.2 Future Shock

One writer during this period who had the most lasting influence on me and many others was Alvin Toffler. He is best known for his book *Future Shock* (Toffler 1970), but I experienced his thought earlier in a piece he wrote for the hardbacked magazine, *Horizon*, titled, "The future as a way of life" (Toffler 1965). I read it also as "the future as a source of identity", and joined him for a while in several futures-oriented activities, including creating the Institute for Alternative Futures in the Washington, DC area, and in considering creating a new political party—the Futures Party—until we realized "the future" has no consituency. It is only specific futures-oriented policies and actions that people are concerned about. Not "the future" in the abstract. Newt Gringrich was a member of that small group, and Clem Bezold directed the Institute for Alternative Futures for forty years with great success.

Toffler suggested that the root of the many discontents of the time was "future shock" which many people experienced because the realty of the future-at-a-later time (i.e., the present) was so catastrophically different from what they had been led to expect it would be. Future shock is similar to culture shock and homesickness. In the case of homesickness, you are eagerly looking forward to leaving your old boring home and heartless parents and going off to camp for a few weeks. You can hardly wait! You arrive at camp breathless with anticipation and wonder....which rapidly drains down into fear, loneliness, dread, a deep desire to return home. You begin moping and crying and wanting to go see your dog NOW. You phone your parents but they say they won't come and get you; you have to stick it out. At the end you are both sad and happy to leave camp and be welcomed and comforted by your boring parents and loyal dog. You are finally back to normal, where you belong, again.

In the case of culture shock, the phenomenon is similar—eager anticipation, initial delight and wonder at all you see in the strange new land which soon wears off when you learn that nothing is as it seems. That foreigners are not exotic, they're neurotic, crazy, with bad food, bad breath, surly manners, unintelligible grunts. You want to go home NOW! If you were enslaved, indentured, marooned, castaway, a refugee, you may be stuck and have to grin and bear it, while if you are traveling abroad freely, on vacation, perhaps, or a "junior year abroad", you can return home where everybody knows your name, and you can understand them when they say it.

Since all life is lived "forward" with our memories, and all education is about the past, what Toffler called "the premature arrival of the future"—future shock—is far more devastating and alienating than homesickness or culture shock. You can't go back to the past in reality, and so you do it in fantasy—you reject a President who says to turn the heat down and wear a sweater to conserve energy, and elect a second-rate actor on an imaginary white horse as president of the United States who declares it to be "Morning In America" and immediately causes the sun to rise by stealing it permanently from the future; transforming the US from the Number One Creditor Nation in the world into the Number One Debtor Nation in three years by fanning the cooling embers of the Cold War and railing against some Evil Empire while ramping up military spending, both of which stole wealth from future generations to sedate current generations. America and the rest of the world fell for the neoliberal shell game from which we have never recovered. The last politician to try to speak truth to citizens, Jimmy Carter, was so thoroughly rebuffed and humiliated by the actor that every politician since has fueled their own fairyland fantasy of the present by continuing to steal wealth from the future in order to reward the 1% while muffling the 99% as well as plunging each succeeding generation deeper and deeper in debt until first brutish neonationalism and then Covid-19 reared up to steal it all and forever until a new kleptocracy can be rigged again perhaps materialized by China, India, and Russia while the US is too saddled with denial and regret to conjure up a future.

## 2.3 Postmodernity

There was a time when I would have had to say something about Foucault about now since his influence on notions of identity across the academic disciplines was so overpowering. But I think that time has passed. It might seem like discussing Christianity without mentioning Christ—or at least Paul—but that might not be a bad idea either.

In the 1990s, the futurist, author, and president of the World Academy of Art and Science, Walter Truett Anderson, wrote several books and articles on "the future of the self". In "The self in global society" (Anderson 1999) he discussed what he called "postmodern psychology". Futurists, feminists, nonwestern thinkers and others were "questioning the traditional model of mental health expressed in Erik Erikson's influential views on how a person ideally grows, gets through his identity crises, avoids the perils of 'identity diffusion', and becomes a stable adult with an assured sense of inner continuity and social sameness. Instead, they are describing the consciousness of contemporary people with such terms as 'multiphrenic', 'protean', and 'decentered'. Feminist theorists...have added a fourth term: 'relational'. They argue that women tend to be more 'context-dependent' or 'relationally-oriented' than men" (which sounds a lot like other-directedness to me). Brain scientists "have abandoned the Cartesian image of a single observer within the brain. Contemporary investigations reveal instead a multicentric brain in which different functions are distributed in complex ways, but not simply located in a single place." (Anderson quotes Michael Gazzaniga as observing that "metaphorically, we humans are more of a sociological entity than a single unified psychological entity". Moreover, "cyberspace is an entirely new kind of social environment, containing a multitude of communities and offering enormous opportunities for self-diversification" (Anderson 1999: 804). To put it mildly!

The idea of the solitary self was revealed to be very ethnocentric—a western pretension: "Anthropologist Clifford Geertz believes 'The Western conception of the person as a bounded, unique, more or less integrated motivational and cognitive universe… is, however incorrigible it may seem to us, a rather peculiar idea within the context of the world's cultures' that may be fading away even in the west" (Anderson 1999: 805).

In 1999, Anderson posited two future scenarios for the western self.

One sees a world of no boundaries where nationalism and ethnocentrism are gone. The diverse religions have merged into one world religion of many amorphous and emergent features. "Governance, society, religion and ecology are all now recognized as art forms" (Anderson 1999: 810–811).

The other future is "back to basics" featuring "a utopian return to communities, ancient ethnic identities, local economies, and traditional spirituality." "But then—to the surprise and great dismay of the idealistic devolutionists—their ideology is seized upon with even greater enthusiasm by religious fundamentalists, political conservatives, ethnic nationalists, racists, and neo-fascist paramilitary groups" (Anderson 1999: 811).

To put it mildly, indeed! We now know which forecast was correct, and that Anderson understood its potentiality well ahead of many people. The world is currently locked in the second future, while, as we will see, more and more individuals and groups are struggling to break free of it and be the ever-evolving human becomings they desire.

## 2.4 Immigrants and Identity

The 19th and 20th Centuries were periods of substantial immigration from Europe to North and South America, Australia, and New Zealand. Europeans immigrated for many specific reasons, some in the hope and expectation that they could create better lives for themselves and their families in these "new worlds" than they could in the places of their birth. Some of these immigrants could and did return to their homelands frequently or permanently, and/or kept close ties with relatives and friends in the communities of their birth. Most of these became citizens of the country to which they moved when it became independent, creating new identities in new nations.

Other immigrants escaped grinding poverty or famine in Europe by signing on as indentured servants for a period of time, while others still left Europe to escape religious or cultural persecution. At the same time significant numbers of people were captured from their homes and forcibly sent to the new lands to live and die as slaves—both completely cut off from their cultures of origin and yet unable to become full creators of the cultures in which they were forced to labor.

None of these new worlds was unpopulated before the immigrants moved in, uninvited and unwelcomed. With a few individual exceptions, the new arrivals systematically sought to remove and exterminate the indigenous people in order fully to control the land and other resources entirely for their own purposes. New immigrants, often from different parts of the world, or from different classes and cultures of the old worlds, sought to immigrate during the second half of the 19th Century, often clashing violently with the earlier immigrants who now saw themselves as the proper owners of the stolen land. People from some specific countries or parts of the world, such as the Chinese, were initally admitted to do specific work. However all Chinese were soon forbidden to immigrate to the US.

The first immigrants to North America came from a few European nations with very similar cultural roots—as well as histories of bloody wars against each other going back many centuries. However the various initial colonies of Europe soon formed two new countries—Canada and the United States—with many similarities as well as some fundamental differences. One of the differences was their attitudes towards the next waves of immigrants—should they assimilate as soon as possible with, and become indistinguishable from, the first immigrants (the analogy of the melting pot where individual ingredients lose their unique features and become part of the whole—as is said to be the case in the United States), or should they both keep and celebrate their old culture while also becoming full participants in the new culture (the analogy of Canada as a rich stew made up of many distinguishable components)? These analogies are not mutually exclusive in practice, and may more truly be caricatures than accurate descriptions. Anti- as well as pro-immigrant factions exist in both countries. Moreover, while people from some cultures and ethnic groups assimilate quickly, others—especially descendants of first people who have lived in North America "forever", as well as former slaves who have lived in the US for many years and have no home culture to which they can return—continue to face massive and pervasive obstacles both to assimilation and to separate but equal cultural identity.

And then there is Mexico, the Caribbean islands, and Central and South America with fundamentally different stories and lessons.

It was not just the Americas and Oceania that were colonized by European nations. Most of the world was formally or effectively parceled out to Europe whereupon every effort was made to extinguish the original cultures and remake them in the image of the colonizing nations. Moreover many of these colonies were cobbled together into single jurisdictions from bits and pieces of distinctive cultural communities without regard for their compatibility with one another, or of the damage done to the original communities when segments of one were placed under the control of one European nation and other portions taken over by other European nations.

The challenges to coherent identity formation were profound during the colonization period. They were devastating when rapid "decolonization" occured after World War Two, and then again when the Soviet Empire disintegrated in the 1990s. Almost all wars and conflicts in the world now find their roots in the ruthless expansion of Europeans into the rest of the world during the 17th-early 20th Centuries, and then their sudden abandonment—though that word also totally distorts the reality of economic and military imperialism which continues to flourish throughout the world.

In short, identity and identity conflict is built into the very warp and woof every aspect of every day life today everywhere.

## 2.5 Identity Contested

One of the drivers of my dissatisfaction with identity is not that systemic anxiety about individual and group identity is relatively new in human history and has gone through (and is still going through) changes of meaning and centrality. Rather, I am distressed that identity is used as a political weapon to force people to embrace and perform certain definitions of self, and to reject and attack other understandings. The very notion of an identity group, or of identity politics, and especially the politics that pits one identity group all for itself and all against others, I find problematic not only because of the intergroup violence that it causes and perpetuates, but also because of the trauma it causes within people who find it difficult to be all for one and all against others; who resist some of the beliefs and actions of their group and admire those of Others. This is especially tragic for people who are somehow part of two or more groups, but are forced to choose only one as their identity.

Who gets to say what one's identity is? May each of us choose the identity we prefer, or must we accept what others say we are when we know that we are not? Just because we accept an assigned identity at one point in our lives are we obliged to stick with it forever even when it diminishes us from becoming?

I wonder about the categories themselves, especially if we acknowledge that they are social inventions and not essential—who/what is a woman? So also with Jews, Blacks, queers, transsexuals, incels....? Who gets to define and act on the definition? The problem I seek to overcome is the continuing use and abuse of these terms since they are almost entirely political and historical, and in no way essential.

Nothing I say here questioning assigned identity and favoring fluid multiple personal and communal identities can or is intended in the slightest to deny the horror and evil of the atrocities that existed and still exist when one set of identities plunders others. Women are not to blame for their oppression nor should they be forced to deny it. Men consciously and continuously oppress them. Jews in no way "deserve" discrimination or extermination. Blacks are not inferior and thus warranting of their enslavement and continuing depredation. How can anyone insist on only one mode of sexual expression—or love—and forcefully ban all others? Are there "disabled" people? Or differently abled? If disabled, why are there obstacles to their abilities? Who determines what is normal? Should abnormals be normalized? Are policies attempting to achieve some kind of atonement and reconciliation for past horrors justified now? Yes. Is affirmative action still needed? Given the political realities of the present, yes. But it doesn't end there.

All identity groups—most certainly including that of a "nation" in the "nation-state system" (another relatively new social invention that solved a problem in Europe several hundred years ago and is the cause of endless problems and the solution to very few now)—are more or less pathological, producing pathological people who engage in pathological behavior against other identity groups as well as frequently against members of their own group by insisting on correcting or condemning the behavior of other group members.

My position here is very well stated by Lennard J. Davis who wrote that "maintaining a category of being just because oppressive people in the past created it so they could exploit a segment of the population, does not make sense. To say that one wants to memorialize that category based on the suffering of people who occupy it makes some sense, but does the memorialization have to take the form of continuing the identity? Even attempts to remake the identity will inevitably end up relying on the categories first used to create the oppression" (Davis 2013: 269).

## 2.6 Identities Don't Identify

Moreover, if the purpose of identity is to encompass all the important aspects of each person's characteristics into a single label, they fail.

If I say "that person is a communist", what do you know about that person? Nothing for certain. You can't even be sure that I believe that person is a communist merely because I utter the statement. It could be a slur, a mistake, a joke on my part, but even if I believe it is true, what do you know now about the person you did not know before? What does "a communist" believe, and what not? How does a communist behave and not behave? Am I saying something positive about the person or something negative? What if that person denies being a communist? What if she denies believing a list of things I insist a communist should believe? What if she has never acted as I insist a communist should act? What if she believes exactly what I say a communist believes and acts as I say a communist should act but still denies she is a communist?

What if I call her a feminist? Or a woman? Or a bitch? Or a man?

Those labels tell me—and you—nothing for sure about the person. If she denies every label I put on her are you to believe me or her? If I say she is a woman and she says she is not, is she displaying "false consciousness", like manual laborers who insist they aren't Marxists and don't believe in Marxist theories, even though Marxists maintain they must, and if they don't, then need to be re-educated—like unemployable coal miners who vote Republican, clearly against their "class interest"?

If dark-skinned people whose ancestors were slaves in America enthusiastically vote for and tithe in support of Trump, are opposed to affirmative action, reject food stamps, decry HBCU, and deny that Black Lives Matter—if they insist that their dark skin, African ancestry, and previous conditions of servitude don't matter to them—is it your duty to help them see the errors of their ways, or can you conclude that not everyone you think is "Black" is Black in the way you think the term means; that, instead, it is your duty to accept people as they wish to be seen, and not as you think they ought to be seen—and ought to want to be seen?

What if a person you say is a woman agrees that she is a woman, as well as a mother, a wife, happy to be protected and corrected by her husband who is a man, a breadwinner, a wife-beater and physically abusive to their children. Yet this woman also claims she is a feminist who cares deeply for and acts sincerely on behalf of women? Is she a feminist or is she pathetically deluded, urgently needing to be either corrected or shunned by True Feminists?

Who gets to define what the labels mean? Much of the conflict in the world is between people with different labels—White versus Black, capitalist versus communist, protestant versus catholic versus jew versus muslim, versus hindu and the rest. However, at least as much and often more bitter conflict is between people who insist that they are White and you are not; that I am Black and you are not; that I am a true protestant and you are not a protestant at all even though you say you are because.....

During the 1950s, many people called and calling themselves "Negroes" and other people called "White" rallied together in the cause of "integration"—striving to co-create a society where people could live and work together harmoniously sharing the fruits of their labor equitably regardless of their "race". "Separate", they said, is not "Equal". We must integrate.

But then powerful voices argued that this was a mistake—just another way for Whites to exploit Colored People. Before we can talk about "integration", we need to understand that Black is Beautiful and that Black Power is able to balance White power, some said. So ways of thinking and goals of action changed, as did proper ways of dressing, speaking, greeting, behaving. For a while there was confusion and resistance. While many embraced it, some Blacks said they did not want to be Black as others were defining it for them. Many Whites who had risked their lives and livelihood in support of integration were shocked when they were told to go back to where they had come from and do their work within the White community. Something similar happened to men who supported "women's liberation" early on and were told to go liberate men so women can get on with their work. This all made sense.

And it didn't.

Labels are sociological statements, not physiological ones. If they have any relation to reality at all, they may be statements of mean, medium, or modal features around what may appear to be a neat bell-shaped curve with two tails, but isn't.

Certainly gender is not and never has been a dichotomy—either male or female. If the term has any meaning at all, a gender distribution would display a complex N-dimensional U-shaped curve with hypothetical hyper femininity at one end and hypothetical hyper masculinity at the other with all sorts of variation in between. There are many more manifestations or indicators of gender than just two—biological, physiological, psychological, social, cultural, political—and they are not fixed. Individuals may flow back and forth, in and around many gender possibilities over their lifespan. "Sometimes you feel like a nut, sometimes you don't", the old ads for Peter Paul *Mounds* and *Almond Joy* used to proclaim.

And there is more to any person's life than either being a nut or not. Each person has many identities—woman, mother, wife, Southern Baptist, Puerto Rican, Democrat, Registered Nurse living in Detroit.... Sometimes the diverse identities of individuals coalesce into groups that function as coherent and powerful communities unified in opposition to everyone else with other mutually exclusive sets of identities and groups—as America is said to be tending now: Republicans are Anti-Democrats, Democrats are Anti-Republicans, and that is all that matters. Independents are just lazy, stupid, or lying. Battle lines are clear—you are either with us or against us.—Hatfield or McCoy. So loyalty tests, purges, traitors, assassinations, and suicides (real and metaphorical) become frequent, bloody, and triumphant. As Stalin's Show Trials, Mao's Great Leap Forward and Cultural Revolution, Hitler's *Sondergerichte*, and Senator Joe McCarthy's Anti-Communist hearings of the US Subcommittee on Investigations each illustrated very well (and impending Woke/Anti-Woke legislative hearings will soon illustrate), there is no end once purification and loyalty to your group are the prime aims of life. There are always evil people to exorcise from the group with extreme prejudice.

However, in reality all the cross-cutting identities are never lined up clearly on one side or the other, even within each individual, much less within each group. It is very hard indeed to be a True woman, a True mother, a True wife, a True Southern Baptist, a True Puerto Rican, a True Democrat and Registered Nurse living in Detroit. The more True Identity is required, the more uncertainty, recrimination, guilt, angst, doubt, hypocrisy, and defection reign, amid battle cries of "inauthenticity". And yet it sometimes seems it is very difficult to be a bishop who loudly and notoriously rails against homosexuality while not also trolling for little boys.

## References


Anderson, Walter Truett. 1999. The self in global society. *Futures* 31: 804–812.

Davis, Lennard J. 2013. The end of identity politics: On disability as an unstable category. In *The disability studies reader*, ed. L.J. Davis. Routledge.

Gleason, Philip. 1983. Identifying identity: A semantic history. *The Journal of American History* 69 (4): 910–931.

Gove, Jennifer, and Stuart Watt. 2004. Identity and gender. In *Questioning identity: Gender, class, identity*, ed. Kath Woodard. London: Routledge.

Hall, Stuart. 1992. The question of cultural identity. In *Modernity and its futures*, eds. Stuart Hall, David Held, and Anthony McGrew. Cambridge: Polity Press.

Kim, Tae-Chang. 1999. Bricolaging a public philosophy for future generations. In *Co-Creating a public philosophy for future generations*, ed. Tae-Chang Kim and James A. Dator. London: Adamantine. https://therisingsky.wordpress.com/2012/10/27/kanji-tip-1-人間-human/.

Toffler, Alvin. 1965.The future as a way of life. *Horizon*. Summer.

Toffler, Alvin. 1970. *Future shock*. New York: Random House.

Yelen, A. 2022. *Poetry* 219(6): 567, March.

# Chapter 3
# The Big Three: Class, Gender, Race

**Abstract** Focusing mainly on the United States, the main markers of identity—one's social class, gender, and race—are discussed, ending with a summary of the argument of the necessity and possibility of moving beyond fixed human beings to endlessly fluid human becomings.

**Keywords** Affirmative action · Black lives · Class · Diversity · Donald · Future generations · Gender · Identity · Population · Race · White

We are each awash in a stream of identities almost without number. Some identities are more pernicious than others, and some are thought to be far more fundamental than others. Class, gender, race are often said to be the big three. These three twirl around in an endless dialectical waltz, sometimes one or two rising to more visibility and provoking more *sturm und drang* than the other one or two. In the US, class seldom gets the recognition it deserves and does get in other countries. This in part may be because it is so utterly un-American to imagine there might classes in the US; it is, shall we say, *déclassé*? We are obliged to recite over and over the All-American creeds that justify an unnecessary WAR for Independence from England and all subsequent gratuitous acts of bloodshed while denying that some men are created far more equal than others, with liberty and justice for a few in the pursuit of happiness via material acquisition. Any mention of the facts of grotesquely-increasing economic (and hence, power) gaps is immediately labeled "Socialism!" and that is the end of the conversation. No True Patriot can be expected hold such a repulsive concept in mind long enough to rebut it much less to use it as a tool for understanding the foundations of identity. Yet social classes do exist in the US even if most people deny being a member of one. (See: "false consciousness" in action!).

Recently, some scholars have sought to resurrect the term "caste" to designate the American system of stratified, birthright inequality, perhaps to separate the phenomena from the (gasp!) Marxist roots of class, but I don't expect caste to gain traction anytime soon. It is a foreign concept for a foreign condition that can't happen here. Family ties don't matter to us! We are all born with the same silver spoon up our nose. Trump is wealthy because he earned every penny of it. Losers aren't, because they didn't, and stupidly paid taxes on what they did earn.

 
J. Dator, *Beyond Identities: Human Becomings in Weirding Worlds*,
Anticipation Science 7, https://doi.org/10.1007/978-3-031-11732-9_3

In the early days of the reign of The Donald, I wrote a kind of tepid "class analysis" of Trump's popularity among America's working and lower middle classes. He presented himself as the kind of unlearned, pussy-grabbing, self-absorbed, raucous, vacuously-religious, fragile, frightened, Good Ole Boy family man that folks in laboring and farming communities know and love too well. On the other hand, his behavior shocked and deeply offended those of the professional classes who struggle mightily to be curious, respectful, empathetic, thoughtful, spiritual, polite, and altogether woke. Trump's hourly, increasingly outrageous lies and tantrums perpetually insulted and provoked them, which perpetually goaded him and his enraptured followers to go ever lower and more obscene. They loved to see their haughty masters convulsively twisting their nappies into wetted knots while he crossed his arms and smirked like *Il Duce.*

To comment on this is to be concerned with class at what might seem to be a superficial level. But it worked because a lot of the most telling differences between the classes are not their differing relationships to the modes of reproduction, as Marxists have it, but to modes of communication, especially to modes expressing humor. Taunts usually sting more than puns and droll wit.

The Donald showed his true class allegiance with all the perks, privileges, and praise he poured upon the rich and superrich, through tax breaks most spectacularly but also by destroying environmental, economic, and cultural rules and regulations that gestured towards the slightest hurdle to unrestrained exploitation and greed. There was a thought at first that Trump would be a populist in policy as well as in rhetoric and behavior, but that thought faded quickly. The words were crudely populist, but the actions were capitalist of the rawest sort: a winning combination, substantially aided by America's carefully-crafted undemocratic structures of governance which irrevocably give much more political weight to some parts of the country than to others.

In any event, no class analysis ever caught on with the media or people. So how about a feminist, queer, or other gendered critique? Wasn't Trump the epitome of the gross, ruthless, privileged, hetero/cisnormative male? Feminists never had a better target: Puerile Patriarchy Personified! And yet, smack in the middle of "Me Too" and LGBTQIA triumphs, Trump and his Trumpettes were undeterred and unscathed, indeed, they were emboldened.

What stuck was race. I suspect one reason we are unable to perceive class or even patriarchy at work in the US is because from the beginning we have seen the world through colored glasses. It is not clear to me that Trump's extravagant racism is more pronounced or putrid than his views and actions on class, gender, disability, and the rest. His public mocking of a disabled reporter; his touting of his perfect scores as a Stable Genius on simple tests of comprehension; his skillful mishandling of the pandemic topped off by a bizarre performance as Superman cum *Il Duce* cum Howard Roark on a White House balcony after returning from Walter Reed Hospital following very special treatments for the Covid-19 virus, and much more were merciless taunts to anyone who is sick, poor, handicapped, in need of assistance and care—he and only he is strong, invincible, alone.

Finally, somehow after one police brutality too many, it seemed Black Lives REALLY Mattered—in spite of centuries of killing, persecution and neglect. Now justice was to be done. Statues were torn down. Apologies proffered. Restitution guaranteed! Well, that didn't last long. I have been through all of this before in the 1950s, 60s, 70s only to see what little, grudging progress there was in both governmental policies and ordinary behavior grind to a halt and reverse from the 1980s onward when The White Man on the White Horse declared that it was Mo(u)rning in America as Voodoo Economics was let loose on the world.

But soft! Will demographics actually be the key to ending racism? For most of human history, the "average human" was a woman living in China and/or India. "Whites" have always been a minority globally. During a brief period in the eighteenth and nineteenth centuries, Europe grew and exported its surplus population to the "New World" and other imperial outposts, and so the ratio of "Whites" to "nonWhites" on the planet approached 50/50 for the first and only time, with narrow White majorities briefly in North America and Oceania. Since the Whites at that time also had superior technologies for more efficient killing and exploiting, and a God who commanded they subdue, disappear, and/or convert all heathens, Whites were able to rule globally and in some countries for a while. They deluded themselves into believing that their dominance was a fitting, natural, and eternal demonstration of their superiority instead of an extraordinary blip in viable births and lifespans due to clean water, sewage removal, and other public health measures happening among some White communities first. But the demographic wave that Whites rode up has been plunging downward for quite some time, while the nonWhite wave continues to rise steeply. If trends continue, ratios and numbers of White people may be vanishingly small by the end of the 21st Century while Black Lives will Matter Big Time. There may be calls then to hold "walks for Whitey"; to "take a White person home for Christmas"; to construct reservations in the badlands where the quaint ways of some White folks may be preserved from total extinction.

Surely this will be the end of racism!

Not likely. First of all, as Garrett Hardin liked to say, "trends are not destiny", and they frequently reverse. Fertility is a fickle lover. Global *depopulation* was viewed as a huge problem in the 1920s and 30s. Commissions were formed nationally and internationally to address the catastrophe. Raymond Pearl, writing in *Natural History of Population* in 1939, said:

> It seems reasonable to conclude, from the data already presented, that the mean annual growth-rate percent for the world population is steadily decreasing at the present time, and during the recent past. In other words, the decline of fertility that has been noted and discussed in an earlier chapter appears not to be exclusively confined to highly 'civilized' countries, where the populations are mostly sophisticated and eager and adept at birth-controlling. It seems rather to be a world-embracing phenomenon--something affecting man as a species (Pearl 1939: 256).

On page 258, Pearl displayed a graph that showed the population of the world from the seventeenth century to 1932. Because of declining fertility, the curve took a typical "S" curve shape, rising rapidly from 1600 to about 1900, and gradually leveling off. The curve showed the population of the world as reaching a little over 2.4 billion in

2000 (when in actuality, it hit 6 billion) with a peak world population of 2.6 billion in 2100 (whereas many demographers expect world population to peak at 11 billion around 2100).

However, when World War II happened and ended, the Babies Boomed worldwide. Changes in sanitation, public health, and agriculture greatly reduced infant and child mortality while extending the average life span. The Number One problem globally suddenly became overpopulation. I believe it still is a major challenge, but people who favor economic growth (and larger numbers of White people) over human and environmental sustainability now are fretting not only about low fertility locally but also about possible global depopulation by the end of the 21st Century.

In the 1990s, some American courts became aware of the racial chasm between court officials and their "clients" in criminal proceedings. The former were largely White and the latter disproportionately of color. So there was a movement towards achieving greater diversity among the personnel of the court by educating and hiring more Black and Hispanic lawyers, judges, and clerks. Strangely, alternative suggestions for diversifying the kinds of people who appear in court instead—or as well—somehow failed to win the day. But if "affirmative action" ever could play a role in correcting racial imbalances, then surely applying it to decisions about who is apprehended and arrested by police and who is sentenced to prison by judges seems perfect, as Wendell Bell pointed out some time ago. Blacks don't commit commensurately more crimes than Whites but Blacks are overwhelmingly more likely to be apprehended, arrested, convicted, and imprisoned than are Whites. There is no way the enormous differences between Black and White statistics here are due to chance, Bell showed. It must be due to something else—personal and systemic racism, perhaps. Correct the imbalance through the honest application of affirmative action criteria at each step and you will likely end up with a more representative sample of criminals in prison. If young Black men are overrepresented and old White women are vastly underrepresented, for example, the sentencing judge should release or parole young Black men and incarcerate old White women (Bell 1983).

Some time ago I was asked to work with several US State judiciaries on "diversity" and other issues in various "Future and the Courts" projects. I recall attending a workshop of the Wisconsin judiciary in Milwaukee. Before the formal meeting began I suggested that prior to looking forward, we look backward. In the 1990s, racial or ethnic friction in Milwaukee was largely between Blacks and Whites, I was told, whereas 100 years earlier it had been somewhat between "Indians" and Whites, and Blacks and Whites, but mainly between German Catholics and Polish Catholics, and between Catholics and Protestants, largely over the scandalous outrage of intermarriages—apparently nonissues now. Might not a similar change in the manifestation of racism appear in the future, I inquired? Don't focus on Blacks and White alone, I urged. What's next in racism or other forms of seclusion?

People of color aren't naturally progressive Democrats. As their social and economic status changes, so might their ideologies and voting behavior. The vanishing of Whites most certainly does not guarantee that progressive policies will flourish. Racism may persist, but the prime victims of racism often change significantly over time. Not long ago in the US, Irish and Italians were regarded as Colored

People, the riff-raff of slum gangs and grifters, despised by the original noble White Anglo-Saxon "settlers". Now Irish and Italians have dissolved into the White establishment trying to stem the flood of raping immigrants and predatory Blacks. To be sure, some racist, ethnic, and cultural conflicts persist over millennia, such as Jews versus Christians versus Muslims versus Hindus, but partly because these religions base their identity largely on indelible conflicts that allegedly happened long, long ago, instead of on seeking to create mutually preferred futures.

If we don't work to avoid it, not only will other old forms of racism continue but also new targets of racial and cultural conflicts will arise. Some of us have been tracking racism against robots and other mobile artilects for many years. This may seem to some an irresponsibly-crazy thing to worry about, but the fear and hate humans manifest against artilects is rising, and more of us need to address it now.

For the sake of harmonious and just futures, we must move beyond identity, from being to becoming.

## 3.1 Beyond Identity

It is quite clear that the change I am seeking is a long-range goal that will not occur overnight, and certainly not by law, fiat, or force. If it emerges and thrives it will be the result of a long, patient process, learning as we go along, from initial actions that lead to others and then still others, with set-backs expected, and errors and excesses corrected.

One of the first steps to be taken is to begin to achieve a balance of emphasis between our concern with the past, present, and futures. While it is necessary for us to seek honesty and atonement for past insults while striving to eliminate those in the present, we should even now also place equal emphasis on considering and exploring future identities—by, among other ways, beginning to "touch fingertips" with future generations.

In a brilliant essay, "To Hold the Grief & the Growth: On Crip Ecologies," Kay Ulanday Barrett writes, "if we were to imagine beyond, we could connect ourselves to our ancestral roots, and really listen, and really give our care in the ways that we are able, what a beauty and what a bounty that would be" (Barrett 2022: 319–320), Someone at *Poetry* was so impressed with this statement that it was printed and included separately on a large card mailed with the magazine.

Well, yes, but in the magazine itself, Barrett immediately follows this familiar sentiment with a quotation: *"The day will come when crip world will be the only world that survived. Crips will do anything to survive and that's what they want to deny when they kill us ... our will to live is greater than your ability to get rid of us."*

*—Maria R. Palacios, Sins Invalid, "We Love Like Barnacles"*. (Barrett 2022: 320. Italics in original).

Ah yes, that is the point—to live our lives for the sake of future lives; to focus not too much on the past or one's condition in the present but rather on the obligations we living owe to those yet to be born whose lives we influence by the way we live our lives today.

By definition, "future generations" are not our own children. Future generations are not living now in the present. They are always becoming. They are yet to be. But since we do inevitably act in ways that impact their lives, we have a profound and confounding ethical obligation to think about how our daily behavior may impact them, and thus we should strive to act responsibly so that future generations will desire to thank rather than chastise us for doing our best to act positively on their behalf, or not.

Futures studies is not about "predicting" "THE" "Future". It is about considering alternative futures and envisioning and striving to create and experience preferred futures, continuously trying out new alternatives and preferences as lessons are learned as time goes by.

In order to help bring a carefully articulated preferred future into reality, one needs to address three things—(1) Identify processes that are moving in the direction of our preferred future that we can utilize and strengthen; (2) Identify processes that are opposed to our preferred future that we can try to change, overcome, neutralize, or recruit; (3) Identify new processes that need to be invented, tested, and used in order to move forward.

## References

Barrett, Kay Ulanday. 2022. To hold the grief & the growth: On Crip ecologies. *Poetry Magazine*, January.

Bell, Wendell. 1983. Bias, probability, and prison populations: A future for affirmative action? *Futurics* 7 (1): 18–25.

Pearl, Raymond. 1939. *The natural history of population*. New York: Oxford University Press.

# Chapter 4
# Pioneers Towards Fluid Identities

**Abstract** A summary of the work of people in the US especially who show in their own lives how to move beyond identities others laid on them to those they prefer—feminist, Black, Confucian, human becomings.

**Keywords** Asian · Authentic · Becoming · Childhood · Christian · Confucian · Ethnicity · Feminists · Filial piety · Golden rule · Hawaiian · Human · Identity · Interracial · Latinx · Passing · Race

The heroic, dangerous, inspiring examples of many people who resist assigned identities and seek new ones has emboldened me to move beyond my own experiences in order to encourage others to create future-oriented identities. This quest is often framed as throwing off identities that they have been expected or required to embrace so that their true identities can live and breathe. Through their lives they show that this can be done, but they also make clear that this is difficult, perilous, contentious, dynamic business indeed, with many initial paths tested, rejected, redefined often very significantly. I am more interested in people who are exploring novel identities—and perpetually fluid identities as well—perhaps recognizing that in a wildly weirding world, fluid identities may best slip-slide through the cracks towards life. They are vanguards from whom we can learn, I believe.

## 4.1 Feminists

In many ways, women set the stage and carried out roles and actions that we will see replicated as other groups come to struggle with defining and fulfilling their identities.

It is common to say that female identity went through four waves:

1. Obtaining for women the right to vote as well as some other political actions in the late 19th/early 20th Centuries.



J. Dator, *Beyond Identities: Human Becomings in Weirding Worlds*,
Anticipation Science 7, https://doi.org/10.1007/978-3-031-11732-9_4

2. Expanding the range of women's rights such as equal pay for equal work, safe and legal abortion, control of their sexuality, and protection against domestic violence, from the 1960s.
3. Further expanding the range of identity and meaning greatly during the 1990s, challenging conventional assumptions of what a woman is, and endeavoring to include issues of race, class, as well as various sexual orientations equally within the feminist sphere.
4. The fourth wave currently is intensely action-oriented while endeavoring to be inclusive of all possible marginalized women and post-women, as the label LGBTQIA+ manifests.

In an article titled, "Feminist theory today," a treasured colleague, Kathy Ferguson, points out three features of feminist theory now: It is "reliably suspicious of dualistic thinking"; it is "generally oriented toward fluid processes of emergence rather than static entities in one-way cause-and-effect relationships'" and it is "a political as well as an intellectual enterprise. It is rooted in and responsible to movements for equality, freedom, and justice."

"Feminist theory," she says, "has generally followed Beauvoir's insight that we are not born, but rather we become, women." This creates a dilemma: "To challenge oppressive power relations, we have to develop our voices. But who are 'we'? For identity-oriented feminisms, the enemy has been essentialism, sometimes to the point that it becomes difficult to speak of biology at all, lest we reinvoke patriarchal 'biology is destiny' bromides." "Yet, articulating a voice requires attention to bodies and locations that are unavoidably material" (Ferguson 2017: 78).

Feminist theory's orientation "toward fluid processes of emergence rather than static entities" is of special importance to my concerns about identity.

Even more directly, Allison Weir persuasively shows that identity is "a source of individual and collective meaning, that enables one to be oneself." But it also is "a source of oppressive constraint, …an obstacle to self-creation and a barrier to freedom" (Weir 2009: 334). Her goal is to reconcile these two, and she does so by arguing that "the quest for freedom and the quest for meaning can be balanced only through a conception of freedom in relationship" (Weir 2009: 335). She agrees with Taylor "that we are essentially dialogical beings. We define our identities through relationships with (and struggles against) others—and ourselves." But there is a big danger here: "This is a communitarian ideal, and it can be a conservative one: it can focus on a tradition, on preserving a past." This "opens us to the danger of entrapment by labels and categories—and worse, by labels and categories not of our own invention, imposed on us by others" (Weir 2013: 1). Nonetheless, she maintains that "the focus on connection can also be an orientation to a future: to alternative, resistant identities, to the creation of new defining communities, to an ideal future, a better, more meaningful life, in a better world" (Weir 2009: 544). We should come to "an understanding of ourselves as active participants in the constitution of our identities, and to an understanding of practices of identification as practices of freedom…of what I call transformative identifications" Weir 2013: 4). "It becomes possible to move beyond the dichotomy of preservation of the past and orientation to the future with a conception of identity in connection to past and future, through reinterpretive preservation and transformative identification" (Weir 2013: 17).

## 4.2 Confucian Human Becomings

Roger Ames, Humanities Chair Professor in the Department of Philosophy of Peking University, China, has written an extraordinary book marshaling strong evidence from a lifetime of scholarly research that I believe supports the positions I am examining here as well. In it, he "argues that the Confucian tradition—particularly the Confucian conception of relationally-constituted persons as 'human becomings'—has an important contribution to make…as we struggle to resolve our current human predicament" (Ames 2021:14). To the extent our overwhelming economic, political, social, and especially environmental challenges are the consequence of humans acting in accordance with ideas about humans, nature, and their interrelationship that derive from the classical Greeks onward, they cannot be resolved within the western paradigm itself. Ames explains, the "most fundamental difference between these two contrasting classical cosmologies is thus the prominence of 'substance' as ontological ground in the classical Greek tradition, and the fluid 'process' orientation of the classical Chinese narrative, defining us as either discrete human 'beings' or as eventful human 'becomings,' respectively. A corollary to the privileging of a formal, unchanging reality over the flux of appearances in the dominant classical Western worldview is the tendency to privilege the discrete and quantitative over the continuous and qualitative" (Ames 2021: 368).

This distinction, and a persistent misunderstanding in the west of the Confucian tradition as an oppressive and rigid set of obligations between hierarchically-aligned individuals, is in part because the Confucian texts were originally translated by Christians who forced Confucian ideas into Christian tropes, Ames points out. This distortion has been perpetuated because many East Asian scholars in the west first learn Confucianism from Christianized sources that cloud their understanding even if they subsequently learn to read the original Chinese documents. "Confucian culture properly understood celebrates the relational values of deference and interdependence; it understands persons as constitutively embedded in and nurtured by unique, transactional patterns of relations. The question is whether a contemporary Confucian ethic that locates moral conduct within a thick and richly textured pattern of family, community, and natural relations can be a force for challenging and possibly changing the world geopolitical and cultural order" (Ames 2021: 401). This question seems to me to be very close to what Kathy Ferguson and Alison Weir were saying about freedom within interrelationships a few paragraphs above.

When Confucius's follower Zigong asked if there is a single word that captures the essence of how one should live, Confucius replied, "deference" (shu 恕), but this is not deference in the most common western sense. Rather, it means literally "putting oneself in the other's place" (Ames 2021: 27). *Shu* is understood in the sense of the "Negative Golden Rule": "do not impose on others what you yourself do not want." The western version typically states, positively, "Do unto other as you would have them do unto you."

Ames says that the "'negative" version is negative "because it does not begin from the assumption that there is some objective and universal standard that can serve as warrant for 'doing unto others as you would have them do to you.' Indeed, to begin from the presumption that there is such a standard and that one has privileged access to it—and furthermore on that basis to assume that one knows the right thing to do to someone else—is at least condescending if not disrespectful. Instead, by assuming the negative version of the Golden Rule, the task remains open and provisional, allowing that deliberation on how to best grow the relationship with this person can only be pursued through a careful consideration of the needs of this specific person within the possibilities of these specific circumstances" (Ames 2021: 341).

Yes! Absolutely right—"this person within these possibilities within these circumstances." Not generalized to any and all. Yes!

I was blown away when I read this passage because years earlier, long before I knew this aspect of Ames' work, I had made the same argument while writing about the ethical obligations present generations have to future generations (Kim and Dator 1999). The western Golden Rule made sense for millennia in small, homogeneous, face-to-face communities, where "everyone" agreed on the rules, and, more importantly, if I hit you, then you (and your family and friends) could and probably would hit me—which I (and my family) did not want. Deeply instilling the positive Golden Rule made complete sense as a way of keeping a village in peace and harmony.

In the globalized but culturally fragmented world we live in now, not everyone has the same preferences. Indeed what is good—or at least not bad—in one culture may be totally forbidden in another. Thus, I said, we must spend some time in careful conversation when we encounter people we do not know in order to learn how to treat them as they wish to be treated. The new global Golden Rule, I suggested should read: "Do unto others as they would have you do unto them"—perhaps even if it strains your own sense of proper behavior.

However there was another big difference for many of us in the 20th and 21st Centuries. In a village, everyone could in fact "get back" at everyone else if they were treated disrespectfully. Not in the globalized world. More seriously yet, some groups and nations—let's say the US—have the military and technological ability, coupled with arrogance and ignorance, so that they have very successfully been able to "do unto others" who are powerless to do back unto them. Until they did—which was the (unlearned) lesson of September 11, 2001 when a group of people from places Americans had been doing unto without restraint suddenly struck back at us with rapier-like finesse and fury, and America and the world changed forever, though not for the better; instead Americans intensified our aggressive attacks on Others within as well as without, as an all-points War on Terror.

The situation with future generations is still more unfair. Unlike most of human history, we now living can and do live our lives in ways that will significantly impact the lives of future generations. And we don't give them a thought when we act. "Why should we?" we may ask. "What has posterity ever done for me"? Nothing, and they cannot "get back" at us to thank us for our consideration or to chastise us for our gross selfishness.

This is yet another reason for moving beyond human beings to human becomings that touch finger tips with all becomings before and as they become.

## 4.3 No Filial Piety

One bone I pick with Ames and Weir is the importance they give to the family as the core locus of all interrelationships. I have asked Ames what he means by "the family" and his replies are not satisfying to me. He refers to the historical, biological, nuclear and extended family systems. While at the present time we may not have anything better, it is very clear to me that many families are dysfunctional all the time and that most are dysfunctional some of the time—often precisely when they are needed most.

It should go without saying that a person, like myself, with no father and not much in the way of a family, has no sense of filial piety whatsoever—whether as something to give or to receive. But even if I had the best father and the most normal family in the world, whatever that might be, I think I would still feel the way I do. I don't think I am the only child in the world who was not only unwanted but whose very presence added unwelcomed pressure and burden on several stressed adults at a particularly stressful time—in my case, the Great Depression and World War II.

Many, maybe most, pregnancies are the by product of other intentions—lust, love, vengeance, violence, duty, habit. While some women may intentionally become pregnant, many of them may wonder whether that was such a good idea when the expectations of their fanciful intentionality are dashed against the rocks of lived reality. Immediately after the dangers and agony of birth itself come, at best, days and nights spent with the wailing and gnashing of gums; of frustration, filth, anxiety. From about age five to twelve, there may be tender, touching, even golden moments of adventure, wonder, and discovery. But then come the teenage years that scar so many who suffer through them whether as subject or object. Many do not survive. If you want to wallow in a world of perpetual guilt and regret, have a child die before you do.

I wonder if it is any different for those who adopt or "artificially" conceive children—are they any more pleased with the lives, misfortunes, and deaths of their decisions? Are adopted or decanted children more consistently loving and loved, any less likely to be abused and abusers? And why do such children often search for their "real" birth mother or sperm donator? How could that possibly matter except as an insult to those who did choose to care and nurture them, however well or poorly, though thick and thin?

Where is the child who chose to be born? Who begged and bargained to be joined in an indelible bond of succor and sacrifice? Why should I feel anything more special towards my mother and father than towards any of the other numerous people who have helped me survive and thrive? Sure, they may have taken care of me even though

I offered scant words of thanks in return. Many people live in a world of fragmented nuclear families in which biological progenitors and their sometime mates are given obligations that are far greater than they deserve or can reasonably be expected to fulfill—they too are often young, immature, selfish babies themselves, struggling to figure out with little support how to survive, much less grow up. How can they be expected to pay proper attention to their accidental and incidental issue? Aside from episodes of care and clinging, parents are more likely to be driven to take pay attention to their offspring, however sprung, from the pressure of cultural mores, morals, or legal requirements than out of love. I appreciate what they do—in my case they did more and better than any three women could have been expected to do—but they no more solicited shows of love on my part than they felt obliged to show me.

"Oh, my pa-pa, to me he was so wonderful," sang Eddie Fisher. Could be. Good for him. I wouldn't be knowing, or caring.

"What's love got to do with it?" Tina Turner rebutted. "What's love but a second-hand emotion" indeed? One reason we are bombarded with love songs is that love is so deeply unnatural, so fleeting, so flawed, so fanciful that we have to have paeans to its magic pounded into our brains. *Eros, phila, storge, agape*—down the emotional sluiceway they flush from the easiest to feel and act on (indeed, the hardest to repress) to the hardest to imagine much less faithfully to express.

## 4.4 Childhood

> The history of childhood is a nightmare from which we have only recently begun to awaken. The further back in history one goes, the lower the level of child care, and the more likely children are to be killed, abandoned, beaten, terrorized, and sexually abused.
>
> That this pattern has not previously been noticed by historians is because serious history has long been considered a record of public not private events. Historians have concentrated so much on the noisy sandbox of history, with its fantastic castles and magnificent battles, that they have generally ignored what is going on in the homes around the playground. And where historians usually look to the sandbox battles of yesterday for the causes of those today, we instead ask how each generation of parents and children creates those issues which are later acted out in the arena of public life (deMause 1974: 1).

Even though Lloyd deMause offers some evidence that some portions of humanity have improved childrearing practices somewhat over the last century (and there has been substantial amelioration in the laws of many countries since deMause wrote), there is good reason to believe that childhood traumas still are a root reason behind the identities and actions of national leaders as well as ordinary people today, most vividly some recent ones—invading innocent countries, craving endless money, begging worshipful adoration, pretending to be strong in order to prove themselves to daddy.

I was a university professor all my professional life, and my office was frequently visited by men and women who blamed their parents for their problems. I don't recall many lavishing praise on their parents for their childrearing skills. Divorce rates are still high and the proportion of people who refuse to marry at all is higher still. The "typical" family in the US is a single woman with two or more children, though "families" of a single person are increasing in frequency too.

It is thus our opportunity and duty to help see that the lives of all children in the futures can be even better by enabling them to form their identity not based on the past, and certainly not dependent on any special obligations to their biological family, by co-creating their individual and collective futures.

While deMause's view certainly has its critics, it has learned supporters as well. Ashis Nandy, of the Centre for the Study of Developing Societies in New Delhi, India, and a harsh critic of much about western society, agreed with deMause that "the tradition of child care is indeed the tradition of neglect, torture and infanticide. So-called parental care and education have often been a cover for wide-spread social and psychological exploitation of children." The evidence does "suggest that mankind has progressed towards better treatment of children and that modern societies have been kinder to children than traditional societies" (Nandy 1984).

*Goal Number 9* of the 2002 UN resolution on a "World Fit for Children" states:

> Children and adolescents are resourceful citizens capable of helping to build a better future for all. We must respect their right to express themselves and to participate in all matters affecting them, in accordance with their age and maturity.

This goal is also a hallmark of deMause's conclusions. DeMause says that there have been six eras of child care throughout history. We have only recently moved into the sixth era, which he calls the "helping mode":

> The helping mode involves the proposition that the child knows better than the parent what it needs at each stage of its life, and fully involves both parents in the child's life as they work to empathize with and fulfill its expanding and particular needs. There is no attempt at all to discipline or form 'habits'. Children are neither struck nor scolded, and are apologized to if yelled at under stress. This helping mode involves an enormous amount of time, energy and discussion on the part of both parents, especially in the first six years, for helping a child reach its daily goals means continually responding to it, playing with it, tolerating its regressions, being its servant rather than the other way around, interpreting its emotional conflicts, and providing the objects specific to its evolving interests. Few parents have yet consistently attempted this kind of child care (deMause 1974: 52–53).

I believe that since deMause wrote these words there has indeed been a significant change in childhood and parenthood practices within some families (though not in others) along the lines he prophesized. These changes have been mandated by substantial modifications in family law in some parts of the world. The adage, "spare the rod and spoil the child" that I heard and saw enacted repeatedly in my childhood (though never on me), is now rightly judged child abuse. The rod wielders may have to pay for it in courts of law.

You may be shrieking and pulling your hair out by this time. I leave it to you to decide whether or not we are surrounded by spoiled adults as a consequence of just this kind of cockamamie balderdash. The complaints I hear from older generations about Millennials and Gen Xers suggest they might have profited from some measured lashes on bare buttocks. I disagree. I'll take a planet full of spoiled brats anytime over a planet ruled by child-abused psychopaths, such as most politicians and many CEOs seem to be (Lasswell 1930/1986).

## 4.5 Race and Ethnicity

Peter Apo asked "How should we define Hawaiian in the 21st Century?" (Apo 2021). He pointed out that "Hawaiian" is an English and not a Hawaiian word. It originally designated citizens, without regard to their ethnicity, of the nation called Hawaii that was created between 1782 and 1810 by the forceful unification of all the islands of the archipelago by Kamehameha I. The indigenous population of the Hawaiian Islands in 1776 may have been as small as 200,000 or as large as 800,000 when the British explorer Captain James Cook first encountered them. However, by 1900, through direct violence, imported disease, and deculturalization, the native population was reduced to fewer than 40,000. The number of citizens of the US State of Hawaii with any Hawaiian blood has grown over the years, but they are still very much a minority of perhaps 12% among the total population. Some estimates say that there are only about 8000 "pure" Hawaiians.

It is worthwhile commenting here that in regards to "Hawaiian" as an identity ascribed to them by others rather than one used by the indigenous people of Hawaii to identify themselves, many refer to themselves as the *Kanaka Maoli* which can be understood simply to mean "a human being" or "people of the land". Once the term *Kanaka* was used disparagingly but many native Hawaiians today use it as a term of pride. In a review of two books about the Aztecs, J. H. Elliott comments, "Both authors have difficulties not only with 'empire' but also with 'Aztec', which is a highly questionable term. The inhabitants of Tenochtitlan and surrounding regions that recognized their dominance were technically Mexica, but as far as is known, the Mexica, along with other peoples of central Mexico, never identified hemselves as 'Aztecs'. Irrespective of their geographical location and political status, each ethnic or social group referred to itself when dealing with outsides and others as 'we people here.' To avoid inconvenience and make the nature of their topic clear to nonspecialists, Berdan and Townsend tend to fall back, with obvious misgiving, on 'Aztec'" (Elliott 2021: 32). This may be true of all first people almost everywhere.

Apo says that what it means to be a Hawaiian has evolved over the years and at the present time is signified by one or more of the following indicators: "Hawaiian by ancestry" in two categories—(1) those with 50% Hawaiian blood quantum who qualify by law to apply for certain Hawaiian homestead land, and (2) those with any amount of Hawaiian blood quantum that qualifies them for certain federal entitlement programs. Because of quirks in their enabling legislation, the first group is designated "native Hawaiians" and the second is designated "Native Hawaiians". Apo says a second indicator of being Hawaiian is "by what you look like"—once a very clear indicator, but, because of genocide and the frequency of ethnic out-marriage, not at all clear now in many instances. "Hawaiian by cultural credentials" applies to a very small group of people with no Hawaiian blood who have more than mastered the language and arts in ways that contribute to the preservation, understanding, and spread of Hawaiian language and culture. "Hawaiians by lifestyle" designates a larger group of people—many from families that have lived in Hawaii for generations—who engage in contemporary Hawaiian cultural practices. Finally, Apo says, are "Hawaiians by living Aloha": "Aloha is a verb. Aloha is to live Hawaiian by taking responsibility for each other, for all living things, for our places, for ensuring a life of peace, dignity and hope for the people of Hawaii and the future of our children."

Most importantly for my focus here is Apo's statement, "It's my belief that it is not for others to define you. You are who you believe you are." I would guess that many Hawaiians by any definition reject that statement, perhaps calling it "cultural appropriation" (thus perhaps rejecting "Hawaiians by lifestyle" as well). But it may well be a more futures-oriented statement than even Apo imagines.

## 4.6 "Passing": Then, Now, Tomorrow

One summer in the early 1970s, I taught a futures class for high school civics teachers in Hawaii. I used the occasion to prepare a questionnaire about the futures with the teachers for them to administer to their classes in the fall. While the substantive results were unexceptional, I was stunned by the replies certain students gave to the routine question about their race. Specifically, the answers given by the students from Kamehameha Schools, where one must be a native Hawaiian to be admitted to the school.

In fact, most students at Kamehameha did not say they were Hawaiian, but chose some other ethnicity that they felt better defined them.

However, students from the public schools of Waianae overwhelmingly said they were Hawaiian.

I feel certain that if I were to administer a similar questionnaire to students at Kamehameha now (not using the term "race" but perhaps "ethnicity"), all would say they were Hawaiian without hesitation. The change I believe is striking perhaps because of cultural pride, affirmative action, and financial assistance, none of which existed for most young Hawaiians in the 1970s.

America has a long history not only of people of color who resist assimilation into a dominant White culture, but also of people to the contrary who try to "pass" as White, even though they have little "White blood" and were born or raised in nonWhite communities. Some of America's race laws once declared that anyone with one drop of "Negro blood" was Negro (which always struck me more as affirming the overwhelming potency of Black people compared to Whites than evidence of White superiority).

Henry Louis Gates, Jr, "White Like Me" discusses Anatole Broyard who desperately wanted to be a famous writer, not a black writer. He was born black into a family that was identified as black and identified itself as black. He became White by denying, in his words, all the evidence "conspiring to reduce him to an identity that other people had invented and he had no say in". However, Gates asserts that

"In a system where Whiteness is the default, racelessness is never a possibility. You cannot opt out; you can only opt in". Nonetheless "in an essay entitled 'Growing Up Irrational,' Anatole Broyard wrote, 'I descended from my mother and father. I was extracted from them.' His parents were 'a conspiracy, a plot against society,' as he saw it, but also a source of profound embarrassment" (Gates 1996).

But the tables recently have been turned. Several prominent people proclaiming themselves to be Black or Native Americans have been revealed to be entirely White and accused of living a life of lies and misappropriation. Rachel Doležal presented herself as a Black woman who became the local N.A.A.C.P. president in Spokane, Washington, and an instructor in the Africana Studies program at Eastern Washington University. She published a memoir, "In Full Color: Finding My Place in a Black and White World" (Doležal 2017).

Sarah Viren, "The Native Scholar who wasn't", tells a similar story about Jessica Kung, a historian and an American Jew who chose to live her life as a Black Latina and, then, in considerable detail, Viren discusses the example of Andrea Smith, a White woman who positioned herself as an Oklahoma Cherokee, building a significant reputation as a highly-respected teacher, scholar, author, and activist, specifically in defense of Native American culture and people. She called out people who appropriated Indian imagery and artifacts, especially those falsely claiming actually to be a Native American. Smith attacked White feminists who longed to identify with oppressed native culture, adding "Of course, white 'feminists' want to become only partly Indian. They do not want to be a part of our struggles for survival against genocide, and they do not want to fight for treaty rights or an end to substance abuse or sterilization abuse," Smith wrote in *Ms Magazine* in 1991. Smith became, "an icon of Native American feminism," and aligned herself with many of the most famous feminist and native scholars.

However doubts about her authenticity were raised almost from the beginning, and by 2006 her authenticity was being very seriously challenged. Nothing has dethroned her. Even though she did not produce any evidence of being enrolled in a tribe, or of having Native American parents or grandparents, she continues to teach, write, and speak for indigenous communities. And she continues to identify publicly and proudly as Cherokee even though it is clear that, by the current rules of the game, she is not.

Viren mentions briefly also the case of Ward Churchill in the mid 2000s whose story is similar to that of Smith, including still insisting his Native American authenticity. She quotes John Stevenson, a professor at the University of Colorado, Boulder, as saying "If Ward proved anything, he proved that if you wanted to say you were XYZ, the way you do it is keep saying that and don't apologize."

By September 2020, "a series of white people who had been masquerading in their fields over the years as Black, Latino or Indigenous—six in academia alone by the year's end"—were outed. Viren discusses them briefly, and is generously nonjudgmental. Why not, if identity is indeed a social construct and not something written indelibly in genes or genealogy? Rather, "the meaning and the consequences of our individual racial identities are largely determined by the collective" (Viren 2021).

But what collective? Or what members of the collective are decisive? In the case of "Andrea Smith, the majority of 'others' still saw her as Cherokee—even though Cherokee officials and some Native scholars said she wasn't".

Rebecca Tuvel wrote a thoughtful and courageous piece, "In Defense of Transracialism" (Tuvel 2017) that sought to apply lessons learned from gender identity to the Doležal controversy. Acknowledging an intellectual debt to Susan Stryker and Sally Haslanger, Tuvel argued that "since we should accept transgender individuals' decisions to change sexes, we should also accept transracial individuals' decisions to change races." "[I]f some individuals genuinely feel like or identify as a members of a race other than the one assigned to them at birth...we should accept their decision to change races."

To say that her position provoked a great deal of anger and anguish is to put it mildly. An issue of the journal *Philosophy Today* [62(1) 2018] was devoted to articles critical of it. In her rebuttal she wrote, "although I knew my article's thesis was controversial, I could not have anticipated the response that ensued. Amidst online condemnation of my article, over 800 academics signed a letter calling for its retraction. *Hypatia*'s associate editorial board subsequently apologized for its publication. Feminist colleagues and academics discussed and speculated about various aspects of my identity online, attacked me personally, and accused me of violence. I was called 'racist,' 'transphobic,' a 'TERF,' a 'disgusting person,' 'Becky,' and 'Rebecky Tuvel'" (Tuvel 2017: 74).

In the rest of her rebuttal, Tuvel offers what appear to me to be stout defenses of her position, and concludes, "I realize that my proposal would modify the way we currently understand race, but I see it as an ameliorative proposal that could help pave the way toward a more accepting and inclusive society" (Tuvel 2017: 84). I agree.

Tuvel's position was expanded and extended by Ann Morning, "Kaleidoscope: contested identities and new forms of race membership" (Morning 2018): "This article argues that in the early-twenty-first century, claims of race-group membership are being complicated by technological developments in genetics and in cosmetics, as well as by new respect for subjective self-identification. As a result, there are more paths than ever to claiming and demonstrating racial belonging. In particular, I suggest that four new types of race-group member are emerging: *genetic, cosmetic, emotive, and constructed* [emphasis added]. Should these types come to be widely accepted as genuine race members, racial groups will become more heterogeneous, resembling kaleidoscopic arrays of core and peripheral members who differ in terms of how many qualifications for belonging they may legitimately claim." Morning gives extensive examples of and consideration to each of the four new types of race-group membership.

Interracial marriages are more common in the US now than they used to be, but are still by no means widely accepted, yet. A Pew Research Report for 2017 began, "In 1967, when miscegenation laws were overturned in the United States, 3% of all newlyweds were married to someone of a different race or ethnicity. Since then, intermarriage rates have steadily climbed. By 1980, the share of intermarried newlyweds had about doubled to 7%. And by 2015 the number had risen to 17%."

Census data shows that a significant numbers of Americans change their reported membership from census to census, and that the furor surrounding Doležal would have been a quite different issue in parts of Brazil which is far more racially diverse than is the US. Morning quotes Chinyere Osuji: "[In Brazil] I encountered several whites married to blacks whom everyone knew as white, who had white physical characteristics, yet found their own white identity problematic. They did not identify with white people. Several of these whites talked about the vibrancy of black cultural expressions that spoke to them and denigrated the discrimination against people with darker hues. White wives spoke to me about their love of black hairdos (including my own short dreadlocks) and Africa. They also loved black men." "If Doležal lived in Brazil, Osuji argues, her feelings and self-identification would be unremarkable; she would be 'just another frustrated black woman'", Morning comments (Morning 2018: 1065). Indeed, that is true in many parts of the world other than Brazil. It is the US and a few similar countries that are excessively obsessed with race and racial purity.

Nonetheless, Morning cites evidence from studies of young biracial Black/White persons in the US "as holding what they called 'transcendent' racial identities." "These individuals did not personally use race as a construct to understand the social world or their relative place in it. This does not mean they claimed 'colour-blindness' in some simplistic fashion; these interviewees were well aware of the racial categories in use around them, and some even reported experiences with racial discrimination. However, they understood racial labels as 'biologically baseless' creations imposed on them by society, rather than as springing from their own characteristics or integral to their sense of self" (Morning 2018: 1066).

"Can there be authentic and inauthentic race identities? If so, how can we tell which is which? My argument is not that there can be no consideration of what is genuine when it comes to race, or even of what is 'true' in terms of widespread social consensus. But it requires thinking very hard about what would constitute racial authenticity and veracity, rather than assuming it to be obvious—let alone that there are clearcut racial in-groups and out-groups, undergirded by fixed identities," Morning concludes (Morning 2018: 1069). These are questions central to my quest as well and relevant to all identities, not just race, or gender.

According to Louis Chude-Sokei, Ijeoma Oluo says she "was about 10 when I found out that my whole life I'd been saying my name wrong. A friend of my father's—an 'uncle'—had come to town, and my white mom had dressed us up for the occasion in traditional Nigerian dress. My top and wrap skirt were of a gorgeous orange- and red-printed fabric, hand-sewn by a woman from my father's village in Rivers State. But when this uncle asked me my name, I embarrassed myself and my family by mispronouncing it 'Joma.' 'That is not your name,' he replied. 'Your name is Ijeoma. You have to know how to say your name. It is a very good Nigerian name.' Suddenly my clothing felt tight and uncomfortable, as if my uncle could see that none of this—the clothing or the name—fit me. To this day, when people ask me how to pronounce my name, part of me knows that no matter how much I've practiced, I still don't say it right. It is a good Nigerian name, and my father was a good Nigerian, while I am floating in this space just outside" (Chude-Sokei 2021: 16).

Alan Gilbert discussed Tommy Pico's poetry in "Refuse to settle". In "The nature poem", Pico writes, "I'm a weirdo NDN faggot" and "I'M FROM THE KUMEYAAY NATION, but I don't want to be an identity or a belief or a feedbag. I wanna b/me," with, inserts Gilbert, the freedom "of creating an identity as much as having one imposed, of being able to slip away". Gilbert then quotes Gloria Anzaldúa who writes of "border people, especially artists, [who] live in a state of 'nepantla'...the Náhuatl word for an in-between state, that uncertain terrain one crosses when moving from one place to another, when changing from one class, race, or sexual position to another, when traveling from the present identity into a new identity." Later, Anzaldúa writes of the "New Mestiza Nation" where "different elements in pluralized identities are emphasized depending on the context; in turn, the fluidity of identity puts categories and labels in flux—and at risk" (Gilbert 2020).

For years, many people in the US who were identified as Hispanic were staunch, dependable Democrats. Recently certain people in Florida who had fled Castro's Cuba surprised many analysts by turning out for Trump. The term "Latinx" was invented to describe both groups. However, Thomas Chatterton Williams quotes Ruben Gallego as objecting, "When [Latinx] is used I feel someone is taking away some of my culture." "Instead of trying to understand my culture they decided to change it to fit their perspective." "In my own experience, whenever I've tried to make the point that racial groups are not and cannot possibly be monolithic, I've been accused (often by white progressives) of proximity to whiteness, of having lost touch with authentic marginalized reality. In that case, there seem to be significant numbers of black, Latino, and Asian voters who have lost touch alongside me" (Williams 2021).

Recently, Anti-Asian rhetoric and violence has increased in the US, and yet how to identify the recipients of the hate is problematic (including the term I just used, "Asian"). Emily Couch shows how problematic it can be. "East Asian people did not become as central to Western conceptions of race as Blackness, nor was there an Asian equivalent of the one-drop rule in the United States that defined anyone with any amount of African ancestry as Black. That creates further confusions over mixed-race people; Blackness is often inherited in the United States, but "being Asian" is a matter of fervent debate." She notes that the 1900 Census listed the ethnic categories as "white, Black, Chinese, Japanese, and American Indian" while in "Britain, Asian on its own usually refers not to people of East Asian descent but of South Asians—India, Pakistan, Bangladesh, and Sri Lanka."

Couch states "I was adopted from China as a baby by white British and American parents and have spent the majority of my life in primarily white spaces. I have no connection to the Chinese diaspora in either the United States or the United Kingdom and speak neither Mandarin nor Cantonese." She adds that her life is by no means rare, and that "My experience as a Chinese adoptee raised by middle-class white British and American parents is different from that of a working-class Thai woman who works in a takeout restaurant, which, in turn, is different from that of a financially successful Indian American who works in Silicon Valley." Nonetheless, "I am very much not part of any 'Asian community,' yet I, too, experience anti-Asian racism."

She concludes forlornly, "The insistent use of the term 'Asian community/communities'—no matter how benevolent the intent—reinforces this othering of Asian people and disregards the experiences of people who do not, or cannot, identify with this collective identity" (Couch 2021).

Sandi Tan, a Singaporean-American film maker, points out that "the very notion of identifying as an Asian-American, a political term coined in the late 1960s that encompasses a practically borderless stretch of peoples, can be of vague consequence. 'I identify as me'" (Yu 2021).

Pat Mora wonders, in part,

> Can I be the only me?
>
> Our earth: so much beauty, hate,
> goodness, greed.
>
> "Study. Cool the climate," advises my teacher.
> "Grow peace."
>
> Can I be the only me,
> become all my unique complexity? (Mora 2021)

In "The long awakening of Adrienne Rich", Maggie Doherty, relying on Hilary Holladay inquires, "Which of these women was the real Rich? The dutiful daughter, the star undergrad, the excellent cook? Or the political poet who used every platform she had—and she had many—to criticize violence in all its forms?" Doherty adds that "Rich never felt she had a 'definitive identity,' and that 'the absence of a fully knowable self'—a 'wound, in Holladay's words—spurred her on, to both self-discovery and creative success'."

But then Doherty ends on what to me is a disappointing but all too common note: "The search for the real Adrienne Rich is a tempting biographical task. But it suggests a curious conception of the self, as something prior to and apart from the social conditions that produce it. The ways one is raised and educated, the language one learns, the stories to which one has access: all these create and constrict the self" (Doherty 2020). It is exactly the concept of self (or identity) as something that arises not only from "the ways one is raised and educated, the language one learns" but also and primarily from the stories one makes up, the fantasies one imagines, the journeys of exploration that one takes with others for an entire lifetime that is the basis of my argument here.

In her review of *Whereabouts* by Jhumpa Lahiri, Sigrid Nunez wrote, "The effect of this upbringing on Lahiri was to make her feel less blessed by being multiculturally enriched than disoriented, as if she were 'from nowhere.' She had two languages, Bengali (her only language until the age of four) and English, but neither could be called truly hers, for, as she puts it, she had not chosen one or the other; rather each had been 'imposed' on her—an unusual way of seeing languages acquired naturally in early life that some might find puzzling. After all, as there is no such thing as a choice of mother tongue, there is no such thing as a choice of motherland, or for that matter, of a mother. But does that mean they are imposed? Still, Lahiri yearned for another language, one that would be unquestionably hers precisely because she had chosen it."

I of course find nothing "puzzling" in Lahiri's feelings—but I do in Nunez' puzzlement. My answer to Nunez' question at the end is, "yes". They were imposed *by others* on Lahiri as part of her (unchosen) identity. Lahiri—and everyone else—should be enabled to "choose" her mother(s), and certainly her motherland!

After publishing four books, in English, to "immense critical and commercial success," Lahiri chose to write and publish her next in Italian, a language entirely "foreign" to her that she had to learn for herself, with the title, *In Alter Parole* (In Other Words) (Nunez 2021).

If one's identity is closely bounded by one's mother tongue, are all those who write poems, songs, and stories in a "foreign language"—in English, for example—engaging in verboten cultural appropriation?

The famous futurist Sohail Inayatullah, is a striking example of a human becoming at home in many cultures, languages, and identities—past, present and futures. As he makes clear, "The image of many selves in one does not necessarily mean, however, a commodified spatially shallow self. There can be authenticity. One can be many people; one can through travel, marriage, and other episodes in one's life stages, learn about other cultures and include them in one's selves, creating a pluralistic self. There are successful stories of this—but the success of incorporating the many is based not only on lightness but on depth. It is undergoing the struggle of learning about the other, the light and heaviness of their selves" (Inayatullah 1999).

As John Wilkinson says, "If you're interested in explorations of identity, be warned that this is anti-identity poetry as well as radically anti-individualistic" (Wilkinson 2020).

Nikki Giovanni was asked, "Generally speaking, does group identification strike you as a limited way of thinking about what it means to be a person?" She replied, "I sincerely— and I mean no disrespect— think it's a stupid way" (Marchese 2021).

## References

Ames, Roger T. 2021. *Human becomings-theorizing persons for confucian role ethics*. New York: SUNY Press.

Apo, Peter. 2021. How should we define Hawaiian in the 21st century? *Honolulu Civil Beat*, June 19.

Chude-Sokei, Louis. 2021. Floating in a most peculiar way. *The New York Times Review*, March 21.

Couch. Emily. 2021. We don't have the words to fight anti-asian racism. https://foreignpolicy.com/2021/04/07/anti-asian-racism-black-lives-matter-racial-justice//.

deMause, Lloyd. 1974. Chapter one. In *The history of childhood*, ed. Lloyd deMause. New York: Psychohistory Press.
Doherty, Maggie. 2020. The long awakening of Adrienne Rich. *The New Yorker*, November 30.
Doležal, Rachel. 2017. In *Full Color: Finding My Place in a Black and White World*. Dallas, Texas: Ben Bella Books.
Elliott, J. H. 2021. Mastering the glyphs. *The New York Review of Books*, December 2.
Ferguson, Kathy. 2017. Feminist theory today. In *Annual review of political science*, 20.
Gates, Jr, Henry Louis. 1996. White like me. *The New Yorker,* June 17.
Gilbert, Alan. 2020. Refuse to settle. *Poetry* 217(3), December.
Inayatullah, Sohail. 1999. A depth approach to the futures of the self. *Futures* 31.
Kim, Tae-Chang, and James A. Dator, eds. 1999. *Co-creating a public philosophy for future generations*. London: Adamantine Press.
Lasswell, Harold D. 1930/1986. *Psychopathology and politics*. Chicago: University of Chicago Press.
Marchese, David. 2021. Nikki Giovanni has made peace with her hate. *The New York Times Magazine*, December 26.
Mora, Pat, 2021. The only me. *Poetry* 217(6), March.
Morning, Ann. 2018. Kaleidoscope: Contested identities and new forms of race membership. *Ethnic and Racial Studies* 41(6).
Nandy, Ashis. 1984. Reconstructing childhood: A critique of the ideology of adulthood. *Alternatives* 10(3).
Nunez, Sigrid. 2021. Lost, at sea, at odds. *The New York Review of Books*, May 27.
Tuvel, Rebecca. 2017. In defense of transracialism. *Hypatia* 32(2).
Viren, Sarah. 2021. The native scholar who wasn't. https://www.nytimes.com/2021/05/25/magazine/cherokee-native-american-andreasmith
Weir, Allison. 2009. Who are we? Modern identities between Taylor and Foucault. *Philosophy & Social Criticism* 35(5).
Weir, Allison. 2013. *Identities and freedom: Feminist theory between power and connection*. London: Oxford University Press.
Williams, Thomas Chatterton. 2021. Easy chair: Shades of blue. *Harper's Magazine*, February.
Wilkinson, John. 2020. Not to be resolved: William Fuller's "Daybreak". *Poetry* 217(3), December.
Yu, Brandon. 2021. A vision of Asian-American Cinema that questions the very premise. *The New York Times*. https://www.nytimes.com/2021/02/11/movies/asian-american-cinema.html?smid=em-share.

# Chapter 5
# Identity, Hawaii, and Me

**Abstract** A personal reflection on the centrality of ethnic identity in Hawaii, and my experience as a human becoming who desires to be empathetic towards those seeking ethnic clarity but personally just doesn't get it.

**Keywords** Hawaii · Hawaii 2000 · Hawaiian knowledge · Healani · History · Identity · Japan · Political science · Religion · University of Hawaii

I quoted extensively, above, from an article by Peter Apo about how "Hawaiian" should be defined in the 21st Century. He ended his essay by an appeal to "living aloha" as the best definition. I left it at that as I turned to other things, certain that my doing so would annoy, if not outrage, scholars and activists who know that "living aloha" is often used as a way to gloss over the brutality, racism, and indeed genocidal reality of Hawaii's past, as well as its present and likely futures. While Hawaii truly may be "the least racist place in America", that is a pretty low bar. There is no doubt that racism exists in Hawaii, sometimes in forms similar to those of the US mainland, but often not. "Race" may "only" be a social construction, but racism is real, tangible, hurtful, destructive, and beyond denying.

As Gates and Curran put it, "While, biologically speaking, the idea of individual human races with different origins is as farcical as the medieval belief that elves cause hiccups, the social reality of race is undeniable" (Gates and Curran 2022).

Recognition and discussions of racism in Hawaii have existed since the term came into use. But concern substantially spread and deepened in Hawaii recently as it has in many parts of the world, following the usual cycle of initial shock, doubt, regret, confession, repentance with promises of restitution, to a kind of ennui and boredom, to active denial soon re-emerging as renewed, intensified, and indeed exultant racism.

I spent my youth and early adulthood constantly, rootlessly on the move (though initially largely within the US South), leaving old friends and gaining new ones with no regrets or even any thoughts about it. I was constantly reinventing myself as we moved from place to place, shedding old selves sometimes, strengthening them other times, and trying out new ones frequently. I loved doing so! This continued as I lived and worked in Japan for six years and in Canada for two (where my children learned about the American Revolution from a perspective very different from what they had been taught in America). I have experienced life in Hawaii for more than sixty years,

 
J. Dator, *Beyond Identities: Human Becomings in Weirding Worlds*,
Anticipation Science 7, https://doi.org/10.1007/978-3-031-11732-9_5

and made Hawaii my forever home for more than fifty. Though I have continued to globetrot until Covid stopped me in my tracks, when I moved to Hawaii I never again made my home elsewhere. Though I always could imagine living happily in any part of the world I visited, when I discovered Hawaii and the Political Science Department of the University of Hawaii, I knew that I had discovered my paradise not only in terms of weather, but also mainly in terms of cultural diversity and tolerance for weird ideas and behavior.

When I arrived in 1969 to teach futures studies and Japanese politics in the UH Department of Political Science, an amazing two yearlong exercise in what the futurist Alvin Toffler declared to be an example of "Anticipatory Democracy" was just getting underway, called "Hawaii 2000". It was an extraordinarily comprehensive, deep dive into thinking thirty years ahead by a remarkable cross section of Hawaii's people of all ages, classes, genders and cultures. It was supported not only by all the major shakers and movers of the time—in government, commerce, labor, education, culture, religion, media and the rest—but also by an extraordinary sample of the shaken and removed on all of the islands of Hawaii. Activities were conducted in a broad array of modes and media some appealing to certain tastes more than others, but everyone enabled to find some modes and media that satisfied them. Every club, organization, community, school, church had its own "2000" activity focused on itself, and as a contribution to the Statewide project. Special care was taken to be sure that members from all of the age and ethnic groups throughout the State participated in island-specific activities as well as those that were State-wide. I feel sure that such a vast search for preferred futures for a community has not been duplicated anywhere else in the world to the extent it was carried out here—and it has not been possible to do it again in Hawaii. The entire process was very well documented in the book, *Hawaii 2000*, edited by George Chaplin (editor in chief of the *Honolulu Advertiser*, the morning newspaper) and Glenn Paige, an extraordinary peacenik professor also in the Department of Political Science of the University of Hawaii (Chaplin and Paige 1973).

There are many, many things I would like to share about the Hawaii 2000 process, and the book. The ideological range of beliefs expressed and considered were beyond anything I can imagine happening in the US today. While it was the intent of the organizers to envision and design pathways to preferred futures for Hawaii, there was no attempt to "predict" anything. Nonetheless, in the area of science and technology the book shows we did an extraordinarily good job of anticipating the kinds and impacts of emerging electronic and biological technologies by the year 2000 and beyond.

But we did a horrible job when it came to social and cultural issues. One of the most laughably stupid gaffs was that even though many women from all walks of life participated, "women's liberation" was well underway, discussed in many venues during the process, and recorded in many workshop reports, in the appendix to the book where all the participants, great and small, are listed, all married women were identified by their husband's name: "Mrs. John Kealoha" and not "Leilani Kealoha"!

Wait! Isn't that exactly the kind of identity change I am promoting? I wonder. If losing your identity and assuming that of your husband is eagerly sought, as I suppose it is or was for some women, isn't it an identity lost for all? I was pleased that Rosemary kept her "maiden" name after we married, as did our two daughters when they partnered. But their "family name" derived from me (and not their mother). And it wasn't chosen freely by them. It did serve them well through divorces, however, when they could in fact have changed their names to anything they wanted. When we moved to Canada for a two-year stay my older daughter did change her first name from what her mother and I had given her to something she felt fit her better.

But it is not amusing that the Hawaii 2000 process got the future of Hawaiians and Hawaiian culture so wrong. I can plead ignorance since I was fresh off the boat. What did I know? But in fact, everyone in the know at the time assured me that all trends showed that Hawaiian culture, language, and people were dying out so that by the Year 2000 all that would be left would be the glorious ethnically-mixed "Golden People of the Pacific" whose Aloha Spirit would be a mixture of Hawaiian culture, Japanese rural culture, and US Southern Hospitality. Even though a stately Hawaiian woman, Pilahi Paki, rose spontaneously during the last plenary session of the Hawaii 2000 conference, held in Kennedy Theater on the UH campus, and graciously revealed to the hushed audience (with many weeping) the meaning of each of the letters of A L O H A as she said her ancestors had handed it down to her, we took that as an epitaph and not a prophecy.

The State had not yet removed laws that criminalized certain Hawaiian religious and cultural practices. Many of the most prominent Hawaiians involved in the Hawaii 2000 process did not speak Hawaiian—they had been forbidden to do so by the educational system, and many said their parents, who could still speak Hawaiian but did not, punished them if they uttered a word of Hawaiian.

They knew the *hapa-haole* songs and dances designed for tourists, but hula in its original form and intent was only slowly being revived. Some Hawaiians and locals began voicing memories and hopes in the Hawaiian language as they sang with friends and neighbors in their garages and yards on weekends and evenings, accompanied by ukulele or steel guitars. A renaissance was underway! And we missed it entirely.

Nonetheless, because I had been thrown so quickly and fully into the Hawaii 2000 process, I had been given an understanding of the many cultures and peoples of the islands—especially of their hopes and fears for the futures—that was in some ways unique. I made literally hundreds of speeches and engaged in countless discussions with people of all ages, classes, and conditions on all parts of all the islands over a span of three years. One of my proudest possessions from the era was a ribbon-dedecked tambourine that a group of hippies on the Big Island tossed to me in gratitude—I think it was gratitude—at the end of one such session. I wrote and produced a televised version of my undergraduate introductory futures course, called "Tune to the Future", on KHET-TV, the local public television station. This was the era of black and white mostly live TV, and KHET was one of only five TV stations broadcasting then, so many people watched it. One out of every four programs was taped, and I would go to some location on one of the islands where I would watch the taped show with students enrolled in the class, and any others who wanted to show up. Afterwards, we would discuss the content, and the futures generally.

Rosemary played on a community soccer team named "He Kinipopo" which can mean, "On the ball" (*Kinipopo* was also a Hawaiian word for baseball). Later, she, one daughter and I joined a Hawaiian canoe club, Healani, and for more than a decade we paddled with them, practicing every day in the fetid Ala Wai drainage canal mauka of Waikiki. My crewmates were all blue-collar workers—mostly police and construction workers—part Hawaiian, and strong, passionate, impressive men. I was the lone *haole* and by far the weakest paddler of the lot. They placed me in the fifth seat, in the back of the canoe just in front of the steersman who was our coach and captain. He tenderly offered me sage, welcomed, helpful advice on how I could improve my paddling: "Puuull you lazy haole fuka! Puuull!" Every haughty know-it-all university professor should seek out similar humbling pedagogical guidance.

There were more than a dozen canoe clubs in the Hawaii Canoe Racing Association of which Healani was one. We were competitive as a Club, but never dominant; maybe 4th overall with some stronger and weaker crews. There was a regatta every Sunday during paddling season held at a different magnificent beach site each week. There were crews of all ages and genders in each club, beginning with grade school age and in essence never ending as the "Master" crew categories got older and older as people kept paddling until, literally, they died.

Those races were something to experience—men and women picked flowers and made lei to give victors and vanquished alike at the end of each race. Food was prepared and shared. Entire families attended the regatta that lasted from daybreak to sunset. Races typically involved courses of increasing length from the shore through the surf to a point offshore and return, perhaps once, perhaps numerous times. One big challenge was to get through the turbulent oncoming surf initially, and then catch a wave coming back in, and not *huli* (turn over). It was partly my job in seat five to see that the *ama* of the outrigger stayed on the water by pushing down on the *'iako* so the canoe did not *huli.*

A second sailing season was set for coastwise and interisland long-distance paddling which was another challenge altogether. While some crews were "iron-man" (or woman, or mixed) it was typical for each canoe to change crews several times during a long race by having an escort motorboat place a fresh set of paddlers in the ocean ahead of the canoe. The paddlers in the canoe would jump out just in time for the new paddlers to rise from the ocean and pull themselves into the seats of the moving canoe so as to keep paddling without losing a stroke—not as easy to do with skill and grace as it might sound. It was never my forte. But there was something hypnotic and cleansing about the repetitive paddling itself of a long-distance race.

Among the most spiritual experiences of my life were when, on rare occasion, while the canoes were all lined up in the deep ocean ready to start a long-distance race, a glide of flying fish would skip, skip, skip across the water in front of the canoes just as the starting horn sounded! And there is nothing in the world like being in the middle of the ocean in a tiny koa canoe in the company of all kinds of flying and swimming fauna under a boundless blue canopy and brisk tradewinds.

Later, our much younger son joined neighborhood teams to play the sports that any young person anywhere in the US might play—baseball, basketball, soccer, football. Unlike his older brother who was a good surfer with many local friends who used our tiny—but well-located—apartment as storage and refuge, and even though we live in one of the world's best spots for body and board surfing, our younger son never showed any interest in or aptitude for surfing.

But this was all to the good since we got to meet more people from more parts of the islands—a mixture of classes and ethnicities along with the family-like culture surrounding team sports. The potlucks featuring the foods of the world were as big a draw as the games. We became friends with people we would never have met without Mack's sports.

All but one of our four children went to Maryknoll Schools instead of public schools. That was totally in conflict with my bleeding-heart liberal sympathies, but I wanted them to receive an ethics-based education that the public schools did not provide. Even though I do not care for many aspects of Roman Catholic doctrine and morality, I deeply admire the Maryknoll Sisters, many of whom labored, suffered and died on behalf of poor and oppressed peasants in Central and South America. The Maryknoll motto, *Noblesse Oblige*, summarizes the core belief of my ethical orientation—to live one's life in the service of others. "To whom much is given, much is expected" is often interpreted in a denigrating, supercilious, elitist way, but dying so others may live, in imitation of Christ, is what many of the Sisters taught and exemplified, and I wanted my children to do the same, as they largely have. To the extent they have, it may also be due to six years of Japanese schooling (which does teach ethics) and examples, as well as being surrounded by the similar ethic of *pono* and *aloha* in Hawaii.

The University of Hawaii was a kind of plantation while I taught there. Most of the teachers were White and most of the students and staff were not. The University had been created in 1907 as a place where capable local students who couldn't afford—or weren't allowed—to go to the US Mainland (as children from the mostly White and Hawaiian upper classes could) were able get a good education. UH from the very beginning was co-educational and open without regard to race at a time when most American universities were neither. However, the original design of the campus and curriculum was under the guidance of scholars from Cornell University in upper New York State, which created some cognitive dissonance from the beginning. Agriculture was a major rationale for Cornell University's creation in 1865 and the University of Hawaii was basically a pretty good cow college until well after statehood in 1959.

When the multiethnic participants in the "Democratic Revolution" returned home—Hawaiian citizens of diverse ethnicity (though predominantly Japanese-Americans) who fought valiantly for the US in the Second World War—they organized politically and replaced the old White Republican political-economic order. Respecting the importance of good education, they determined to re-make the University of Hawaii as "The Harvard of the Pacific". In order to do so quickly, they hired a number of world-class scholars and seeded them among new and existing disciplines, along with some junior faculty of promise (e.g., me)—almost all of whom were White (well, many were Jewish, but Jews are White in Hawaii; there is no lived sense in Hawaii of what being Jew—or Black—means, as there is on the US mainland).

At the time, there was no thought of re-creating UH as the university *of Hawaii.* It was to be a world-class university *in Hawaii* with substantial State funding (One of the stories an early President of the University after the "Democratic Revolution", Harlan Cleveland, loved to tell was about frequently receiving telephone calls from the quiet but revered Governor of Hawaii, John Burns, saying, "Harlan, we want to give you some more money!", and President Cleveland having to reply, "Well, Governor, we really can't use any more money now, thank you"). Those days passed quickly. The idea that UH could be a world class university in a community where tourism and the military were the main sources of income and the population of the State not yet one million people was indeed inspiring, but an utter fantasy. Still, the university (and community generally) was a place where any idea could be—and was—freely expressed, and where new, innovative academic programs were permitted without restraint—and wholly without adequate funding, if any, as well. A million flowers could bloom, but they had to be self-fertilizing.

The motto of the University, engraved on the Founders' Gate proclaims, *Maluna a'e o na lahui a pau ke ola ke kanaka.* However these words might be understood by Hawaiian speakers, the English version ("Above all nations is humanity") seems to be yet another legacy from Cornell where it was the motto of the Cosmopolitan Club formed by international students attending Cornell at the very time the University of Hawaii was being created. From the beginning, UH was designed to help Hawaii's citizens to excel in the world—or the cosmos—and except for agriculture had little or no local orientation. It did not even have a Department of Hawaiian Studies. There was a one-man academic option run by the embodiment of Aloha itself, Abe Piianaia, under the umbrella of Liberal Studies created in 1970, but the Hawaiian Studies Program gained only provisional status in 1979 and permanent standing in 1985 within the College of Social Sciences. The Program was placed in the newly-created School of Hawaiian, Asian and Pacific Studies in 1987 and renamed the Center for Hawaiian Studies. A special building for the Center was completed in 1996. In 2007, the Hawai'inuiākea: School of Hawaiian Knowledge was created, incorporating most academic units focusing on Hawaiian subject matter.

School of Hawaiian Knowledge! A huge conceptual leap, that took a long time maturing.

The 2021 Institutional Report for WASC reaffirmation review lists three themes for UH. The first theme is Becoming a Native Hawaiian Place of Learning.

My Department of Political Science created one of the first Indigenous Politics specialties focusing not primarily on Hawaiians but on indigenous peoples everywhere. From the beginning, it had a marked futures orientation, not dwelling on the past more than necessary to understand its contribution to the present and to spur efforts for the creation of preferred futures.

The full story would also discuss the torturous growth of Hawaiian studies units at the University of Hawaii in Hilo and the various community colleges, as well of the UH Ethnic Studies Program that also began in the late 1960s–early 1970s, celebrating its 50th anniversary in 2021—not to mention the enhancement of programs serving Hawaiian children offered by Kamehameha Schools, founded by the will of Bernice

Pauahi Bishop, the great-granddaughter of Kamehameha I, as well as many programs for Hawaiians funded by the Hawaii State and US Federal governments.

There is absolutely no doubt that the overthrow of the Hawaiian kingdom under Queen Lili'uokalani in 1893 by some businessmen with the support of the US Marines, and the subsequent annexation of Hawaii into the US as a Territory in 1898, as well as the limited options offered when Hawaii's citizen voted on statehood in 1959, were all illegal, unjust, and beneath contempt—that is to say, in keeping with many actions from 1492 to today.

There are now many (often conflicting) voices calling for sovereignty from or within the United States, some seeking an independent nation based on Hawaiian ancestry, or asserting it already exists; others based on whoever will renounce their existing citizenship and pledge loyalty to Hawaii without regard to ethnicity; others seeking a kind of state within a state relation similar to that of many Native American tribes on the US mainland. So far the issues are fluid, contested and seemingly beyond quick and peaceful resolution—or violent resolution for that matter.

In any event, the question of "who is a Hawaiian" and what difference it makes, if any, has grown during my lifetime from being well-hidden if not essentially nonexistent to a major disputed feature of the future of every person living in Hawaii, or who wants to, right up there with climate change, sea level rise, tourism, and food safety—not to mention the grotesque fear of an imminent Chinese/North Korean/Russian/Somebody "Pearl Harbor" attack.

Many university professors at major research universities, like the University of Hawaii at Manoa (UHM), are oriented primarily to research, publication (and, if unavoidable) teaching in some specific academic discipline, and their advancement in that discipline. They usually are not particularly devoted to the university (unless it has a great football team, but Portal Transfers may destroy that loyalty too) and its community where they are temporarily located. Anything that distracts them from their research, publications, and academic advancement within the discipline is a waste of time for them. They are somewhat like professional athletes (now, college athletes as well) who are easily "traded" from one team to another, focusing on being the best player they can be wherever they are without developing deep roots in any specific community.

I probably was somewhat unusual in the amount of time I spent engaging as an ordinary person in the cultural life of the community. Many faculty at UHM are hired to do research and some teaching, and will jump ship at the next offer that allows them to do more and better research somewhere else. But Hawaii also is a roach motel—many academics move in but then find it almost impossible imagine moving out. When I got here, I immediately knew this was my home. Many of my colleagues felt the same and stayed forever because of its unique culture and environment. However, there is a tension between the *haole* professors and the local staff, and between the *haole* professors and the ordinary local citizen. In some ways this is just the usual town-gown conflict between differing lifestyles and interests found in any university environment. But in other ways, it is different because of racial issues.

Even though I had taken a great deal of time to learn about Japanese history and language when I lived there, I did not make a similar commitment to Hawaii—nothing

I learned from the Hawaii 2000 immersion or experienced initially at the University of Hawaii gave me the slightest nudge in that direction. All my inclinations and friends were global and transcultural. It never crossed my mind to learn the language. And by "language" I mean not only the Hawaiian language, but also pidgin, the creole patois that arose during the plantation days when people from many different parts of the world had to learn to live and work together. Pidgin was THE hallmark of being "local" when I arrived in Hawaii, but very few of the UHM professors could speak it even though most of the staff and all local students could and did. The Hawaiian language itself was plucked from an early grave by the same people and pressures that provoked the rest of the renaissance, so that Hawaiian is now the first language for some people who also then learn English later. Pidgin too is enjoying popularity but generations of formal education and the profusion of English-speaking media have encouraged almost everyone to speak a mildly local version of Standard American English.

## 5.1 More Bemused Than Empathetic

While I watch with bemused respect people who I love wrestle with their identity, I have no personal involvement in any identity struggles whatsoever. I acknowledge it, but I don't feel it. My attitude here is similar to my attitude towards religion and religious believers. I am not a believer any more, but I am not an atheist either. All of the different things each religion requires one to believe and do in opposition to the things other religions would require one to do and believe in conjunction all the things one might prefer to do and believe leaves me cold. That is something I also learned in Japan—in part by studying Nichiren Shoshu and the Soka Gakkai, its unusually aggressive Buddhist lay organization, and especially the Komeito, its very successful political party that was wholly unique within Japanese history.

Most Japanese are not religious, and view with mild wonder those who are—especially those who are active proselytizers of one religion or another. Most Japanese feel that real people don't need "God" to tell them what to do and not do. That is what other people are for—family, friends, neighbors, experts, the entire engulfing culture: to guide them. But at the same time, "it's OK to believe in religion if it helps you. I don't care", the average Japan might feel. While Shinto, Buddhism and Christianity (and all the rest) in all possible flavors exist in Japan, and while Japanese might pause and clap reverently when passing a roadside Shinto shrine, don all the trappings of a western White Wedding for their marriage, and be lulled by the comforting drone of Buddhist sutras for their funerals, it is mainly significant pageantry and ceremony. It is not a question of saving their soul, going to heaven, and avoiding the eternal flames of hell for most of them.

I can empathize with that. I still revel in the sights, sounds, and smells of an Anglican High Church Mass. I listen "religiously" to the hour-long broadcast of church music every Sunday morning at 8 AM on Honolulu's excellent public radio network, HPR2, on a program called "With heart and voice". The soaring descants of angelic boy choirs can send me weeping far too easily—I used to be one myself, and fondly recall the evil thrill we felt when we succeeded in making the little old ladies cry.

Religion has not only been a balm and guide for countless lost souls; it has also been, and still is, the rationale behind most of the carnage, brutality, and destruction by wars, laws, and bigotry since the invention of writing enabled "civilization" that produced among other institutions, organized religions with their bibles, creeds, priests, and excommunication in contrast with earlier free-floating spiritualty of oral societies, down to the present day.

I feel pretty much the same way about people passionately arguing over identity. That which so self-righteously unites as self-righteously divides and impels us to conquer.

## 5.2 Stories of the Past

I repeat: None of this is to say that we should ignore the systemic evil underlying all our histories, and whitewash our pasts. Absolutely not. Truth is needed, honesty rewarded, and repentance offered and received.

We need to know history as accurately as possible, yes, but inside each history is other history, and inside that still more histories of exploitation, usurpation, as well as cooperation until we are back to the Pleistocene Epoch and the interactions of *homosapiens, sapiens* with other *homo* of the time—*neanderthal, denisovans, longi* (perhaps; perhaps others also)—whose extinction may be due in part to actions by *homosapiens, sapiens.* Before them are hominids and mammals and each of the myriad evolutionary forms of life that contributed to our existence all the way back to the Big Bang (and whatever came before that if the cosmos is not just one big bang but rather is breathing in and out eternally so our Big Bang was just one big exhale following many big inhales and exhales before and after it).

There is no real beginning to our history. It is one huge, eternal, complex interactive process of cooperation, conflict, adaptation, extinction, and luck. So, we should know all of it, yes. But we should mainly focus our energies on the only dimension of time over which we might have some influence—not the past; not the present (it is too vanishingly thin); but the ever-approaching futures that provide us with endless possibilities to (re)discover ourselves, leave our marks, luxuriate in life, thrive, and die.

## References

Chaplin, George, and Glenn Paige. 1973. *Hawaii 2000: A continuing experiment in anticipatory democracy*. Honolulu: University of Hawaii Press.

Gates Jr., Henry Louis, and Andrew S. Curran. 2022. We need a new language for talking about race. *New York Times*, March 3.

# Chapter 6
# More Pioneers of Fluidity

**Abstract** Resuming the story of people who have found fluid identity in their own lives, as an inspiration for others to do the same: queer, transexual, transracial, disabled, carceral, refugees, and environmental refugees.

**Keywords** Acculturation · Assimilation · Carceral · Census · Dator's second law · Disability · Dismodernism · Environmental refugees · Gender · Handicap · Identity · LGBTQIA+ · Normal · Prison · Queer · Refugees · Transgender · Transracial · Transexual · United Nations

I paused my recitation of people who were pioneers in rejecting identities ascribed to them by others that they did not accept for themselves to say a bit about my own experiences as person of indifferent ethnicity and identity within a beloved community of marvelous identity diversity of all kinds—Hawaii; that while I love being a part of this community and am respectful of those for whom their ethnicity is of major concern, I just don't get it.

Now I wish to continue the story of people who excite me even more profoundly than the ones I mentioned before, inspiring me to expand my quest to encourage more human becomings.

## 6.1 Queer

I have been involved in futures studies academically and professionally since the late 1960s when it emerged along with a host of other "studies"—women's studies, ethnic studies, cultural studies, peace studies, environmental studies and the like. But in spite of enormous demand for experts who can make useful statements about the future, and the existence of very successful programs in a few universities around the world, futures studies never caught on as dramatically as expected at the outset, and as some of the other "studies" and newer ones have.

While engaging deeply with the literature of queer theory, I had a major moment of enlightenment about futures studies.

 
J. Dator, *Beyond Identities: Human Becomings in Weirding Worlds*,
Anticipation Science 7, https://doi.org/10.1007/978-3-031-11732-9_6

David Halperin wrote, "English, history, classics, anthropology, sociology, or religion now have the option of using queer theory…to advance the practice of their disciplines—by "queering" them. The outcome…makes queer theory a game the whole family could play. This has resulted in a paradoxical situation: as queer theory becomes more widely diffused throughout the disciplines, it becomes harder to figure out what's so very queer about it, while lesbian and gay studies, which by contrast would seem to pertain only to lesbians and gay men, looks increasingly backward, identitarian, and outdated" (Halperin 2003).

If, at the very beginning, we had named our field "futures theory", and taught it with a thick French accent, it might now be ensconced in every university in the world, along with feminist theory, queer theory, disabilities theory, critical race theory, and the rest. They—especially queer theory—managed not only to survive but also to become wildly popular with deans, dons, and devotees alike, though of course not with regents and legislators.

*Dator's Second Law of The Futures* states that "in a situation of rapid social and environmental change, any useful statement about the futures should appear to be ridiculous". While I have uttered many ridiculous statements that turned out to be useful (and some not so useful), I was too lacking in foresight and imagination in 1970 to declare that we should label our emerging field "futures theory" because doing so might make us appear more respectable and academic, like queer theory. Somehow, neither the word "theory" nor "queer" came bursting from my lips at the time, to my regret.

The emergence and dominance of queer theory is a wondrous thing to behold. Since my earliest university courses about the futures, I have declared that if there is a truly ridiculous dogma we should get rid of it is the last great dichotomy of Western Civilization that asserts that there are two and only two clearly distinguishable genders, male and female. It was obvious to me then and now that this is false, both biologically and socio-psychologically. My assertions and proofs had no purchase for decades as women worried about becoming women, boys were boys, ethnicities struggled to breathe free, and environmentalists hugged trees and smelled the roses. Then, suddenly, it was no longer queer to be queer. It was almost downright normal. Of course, though dethroned, the old dichotomy is far from dead, and frequently attempts to raise its severed head, but many people have finally discovered that the world is a lot queerer than they imagined.

Without necessarily denying that there is a biological base to gender/sex identities as there is to every aspect of life, more people now emphasize the importance of the volitional, psychological, sociological, ideological, cultural bases instead. As we quoted Kathy Ferguson, "Feminist theory has generally followed Beauvoir's insight that we are not born, but rather we become, women." So also with all other variations of human being.

The 1948 blockbuster book, *Sexual behavior in the human male,* by Alfred Kinsey rocked the world not primarily because of its frank though antiseptic accounts of male heterosexual behavior but rather by his accounts of male homosexuality, masturbation, and other sexual activities. How could this be in this very straightened nation? Nonetheless his message basically was that there are two genders that are expressed in one of three combinations of sexual orientations—heterosexual, homosexual (gay/lesbian), and bisexual. Most early researchers in the fields of gender identities tended to emphasize the old dichotomy while admitting, ignoring, denying or delegitimizing other orientations.

Zoe Kolker and friends say, "Queer was first used as a term meaning strange, odd, or suspicious. Later in the early 1900s, queer or fairy was used as an insult to refer typically to gay men who displayed more feminine behaviors. In the 1920s and 1930s middle-class gay men would refer to themselves as queer to distinguish themselves from fairies who were seen as feminine and lacking status as masculine. In the 1980s the LGBTQ community reclaimed the word as an identity label in order to break boundaries of binary ideas of sexuality and gender/sex."

However, "queer is often seen as a 'non-label'; queer is fluid and changing, and therefore hard to define." "Rather than create socially acceptable forms of non-normative identities, queer theory rejects the idea of assimilation to the norm. Its goals are celebration rather than normalization" (Kolker et al. 2020: 1340).

Hannah McCann and Whitney Monaghan remind us that the "...gender binary as discussed by Butler and others is a distinctly Western construct, and that there is a long history of non-Western alternative gender identities. These include hijra (South Asia), kathoey (Thailand), waria (Indonesia), two-spirit (North America), machi (Chile and Argentina) and many more terms and identifications" (McCann and Monaghan 2020: 175).

Bernadette Marie Calafell and Thomas K. Nakayama similarly emphasize that "Queer theory begins from the notion that identities are not destiny; our identities do not determine who we are, who we become, or how we view the world. Instead identities are constituted and constructed in order to meet particular goals.... Queer theory insists on the constructed aspect of identities, and by so doing it opens the possibility of constructing identities in other ways and for other goals." Transgender activists "call for individuals to be able to choose their own legal and personal gender status, to perform multiple gender expressions, and to have civil rights" (Calafell and Nakayama 2016: 4).

One of the consequences of this proliferation of fluid identities is that it has become almost impossible to use one word or a few letters to describe the orientation: "LGBTQIA+" points to this by adding the plus mark at the end.

In the poem, *James River*, Jenny Johnson concludes,

> I am the air red beneath your fingernails,
>
> the trout poking through river rocks
> remote as any one appendage from the other,
> the shadow of what you asked for turning to husk.
> Don't tell me the body won't turn on you like a corkscrew (Johnson 2013).

## 6.2 Transexual, Transracial....

More and more people are refusing to identify themselves by the old gender labels, and yet, in spite of some elasticity, most governmental policies assume the old two only. This creates a challenge for the most fundamental and official set of statistics for all governments, in the case of the US, the Census taken every ten years. Bonnie Ruberg and Spencer Ruelo address this directly in a study based on interviews with people of varying gender identities. They conclude: "Demographic data commonly imagines identity as fixed, singular, and discrete. However, our findings suggest that, for LGBTQ people, gender and sexual identities are often multiple and in flux. An overwhelming majority of our respondents reported shifting in their understandings of their sexual identities over time. In addition, for many of our respondents, gender identity was made up of overlapping factors, including the relationship between gender and transgender identities. These findings challenge researchers to reconsider how identity is understood as and through data." (Ruberg and Ruelo 2020: 1).

The same is true for ethnic labels as well all the other identities that do indeed coalesce and cleave in a vast jittering stew of tangling images. "Some of the growth by Asians in the 2020 census may be rooted in the fluidity of how some people, particularly multiracial, report their identity on the census form, said Paul Ong, a professor emeritus of urban planning and Asian American studies at UCLA. 'People change their identity from one survey to another, and this is much more prevalent among those who are multiracial or biracial,' Ong said" (Tang and Schneider 2022: A12).

## 6.3 Disability

While I have been more or less content with my body as I have inhabited and experienced it as it morphed along through the ages, there are many parts of it that profoundly disable me. I cannot see in the ultraviolet or infrared. I cannot hear dog whistles or many other high frequency sounds. While my sense of smell seems better than that of many humans, it is puny indeed compared to my dogs. I fervently wish I had wings so I could fly—I so envy birds and bats and bees. I have yearned my entire life for a prehensile tail. I have spent years hoping to visit Mars where wings and a tail would be highly desirable—as would a sturdy shell, like that of a tortoise, into which I could retreat when solar flares were about to strike. As I have aged, my body in fact has tended towards that of a tortoise, my hard, scaly back bending slowly over my softening underbelly. But, alas, no tail or wings have appeared yet.

The person I was told was my father drowned shortly after my first birthday, totally changing the trajectory of my life. I have sometimes wished that he and I could breathe underwater as well as on land. But I can't imagine my life would have been so much better had he lived, so maybe it is just as well.

Of course like many handicapped persons, I can use prostheses to solve some of my problems. I fly on airplanes and I have learned bravely to live with my tailessness as an irremediable impairment. I have worn glasses since I was in my late teens to correct mild hyperopia, but now my long-prophesized macular degeneration is swiftly manifesting itself, so blindness will overtake me if I live too much longer. For years, based on observing when many people's abilities cascade in decline, I told everyone I wanted to die before I reached 87, but, alas, I zoomed past my use-by date, and while death might have granted me my most fervent wish by the time you read this words, I am still *regratefully* alive while I write them. My hearing has always been poor, and it has gotten progressively worse. Hearing aids exacerbate the problem.

You could tell a similar story. More or less.

There is no one who is not impaired or disabled. There is no one who is ideal—the ideal body/person is a Platonic invention that no real person can or should try to attain ("Ideal" is like "utopia" in the sense of being an impossibly perfect "no place" which has lured so many reformers into brutality, killing, and totalitarianism). There might be some people who can be called "normal" in the statistical sense of being the average, whether mean, median, or mode, with half of the rest tending towards some kind of supernormality at the end of one tail and half tending towards some extreme abnormality measured somehow on the other tail. By definition, there are fewer and fewer actual people as we approach the end of each tail of a bell-shaped curve; most people cluster on one side or the other of the average. So "normal" should mean just "average". Most of us seem to be from Lake Wobegon (if it is still permissible to speak of that heartbreakingly woebegotten place) "where all the women are strong, all the men are good-looking, and all the children are above average." To be normal seems to mean that you are as close to the ideal as possible, not merely average, and few among us proudly claim to be abnormal (though more of us should, given the obvious pathologies of all normal people).

Normality, Michael Oliver argues, is a construct imposed on a reality where there is only difference (Cameron 2010: 8) while Colin Cameron declares that "being normal involves a process of *learning to want to act* in the ways that society requires one to act." (Italics added) (Cameron 2010: 7).

All of the identities humans fret about are extremely slippery concepts, but none is slippier than "disability" or "impairment". While many cultures do not, or at least did not, exhibit the categories of ability and impairment that exist now in the west (indeed, some abilities that westerners might view negatively disabling were highly prized or respected in some cultures), ever since "modern" times, with its increasing focus on extracting, making, selling, and buying—especially buying—as the main point of human existence, we have labeled people who could not contribute to keeping the economy growing appropriately as disabled, and either in need of being made "normal" or hidden away so the rest of us can shop till we drop without their distraction.

Lennard J. Davis agrees that "we do have to acknowledge that, unlike race, class, gender, sexual preference, and the like, disability is a relatively new category. Although the category has existed for a long time, its present form as a political and cultural formation has only been around since the 1970s, and has come into some kind of greater visibility since the late 1980s" (Davis 2013a: 263).

Elsewhere, Davis says that we should "focus not so much on the construction of disability as on the construction of normalcy. (T)he 'problem' is not the person with disabilities; the problem is the way that normalcy is constructed to create the 'problem' of the disabled person" (Davis 2013b: 1).

Speaking of disabilities in any general—shall I say, "normal"—way as we speak of other identities is also quite difficult because the official list of disabilities is long and growing. The following disability categories were represented in a sample of articles about disabilities: "stroke-related, sight impairments/blindness, spinal injuries, intellectual disabilities, hearing impairments, mobility related, chronic illness, multiple sclerosis, Down's syndrome, hidden disabilities, psychiatric disabilities, multiple chemical sensitivity, obesity as a disability, spina bifida, autism spectrum, and deafness." (Forber-Pratt et al. 2017: 6). As the populations of most nations of the world age, the number of officially disabled people for whom public or commercial accommodation is required is growing substantially.

Internationally, the World Health Organization estimates that 15% of the world's population lives with some form of disability, of whom 2–4% experience significant difficulties in functioning. In the United States of America, approximately one in five Americans have a recognized disability (Forber-Pratt et al. 2017: 12). Unlike other identities, disability can manifest itself at any point in one's lifespan. One can be born "healthy" and experience a "normal" life until accident or diseases causes one suddenly to become "disabled". Some handicaps can be overcome somewhat by more or less intrusive prostheses. Others cannot. Some are obvious to any observer. Many are "invisible disabilities". Some are physical. Some are mental. Some are unlabeled until they become identified somehow. The psychological impact of each these is often widely different.

Some of the literature on disabilities is written by people with a specific disability and reflects their particular responses as well as the specific history of their disability but fails to express adequately the experience of people with other disabilities. A lot of the literature that is written by "normal" people is often declared to be bogus by disabled people. Creating a community of the disabled is a challenge. Some people with some disabilities refuse to be classified with people with other disabilities. A person with impaired vision may not wish to be considered in the same category as a person with severely limited mental capacity or physical mobility. Many don't consider themselves disabled at all and strive to act "normally".

Identities based on disability, handicap, impairment (some people seek to make clear distinctions between these terms) have gone through the same kind of evolution as have other identities—from fixed, rigid, shameful; to contested, fluid, accepted; to laudable, triumphant and prideful: "Mad Pride" "Disability Pride", "Deaf Pride"—not "hearing loss" but "Deaf Gain"; the expansion of other ways of sensing the world (Bauman and Murray 2013); and "Piss on Pity."

Disabilities among family members are frequently concealed in denial and shame. Jennifer Natalya Fink lauds her "family's rich disability lineage" maintaining that "we should claim our disabled ancestors with pride." "Our disability lineages can only be reclaimed through the stories we uncover. This means conceiving of disability as an identity like being queer, rather than reducing it to a medical condition" (Fink 2022).

For many years, the goal was to make disabled people be as "normal" as possible. Indeed the early eugenics movement in the US and elsewhere strove to remove abnormal people from the gene pool so that everyone would be born normal, or aborted before they could be born if they would not be normal after birth. Infanticide of imperfect babies was not beyond the pale. Those who did their best to act normally in spite of their impairments were widely praised as "good little soldiers" who were fighting hard to be as normal as possible, and ashamed that their mere existence caused discomfort to so many nice normal people.

In the 1960s and 1970s, the very concept of "normality" was challenged, especially concerning mental illnesses. Thomas Szasz famously declared that insanity is a myth (or metaphor) and not an illness in the sense of a typical physical disability. He insisted that no person should be imprisoned by the state (i.e., placed in a mental institution) by virtue of their presumed insanity unless their abnormal behavior had a clearly identified basis in biology.

Colin Cameron observes that "everyday life is constrained by social structures and is at the same time an active process of production which transforms social structures." Identity is "a project to be worked at rather than…a fixed characteristic." By "'coming out' and claiming identity as disabled, people with impairments not only subvert dominant disability discourses but…change the very meaning of disability" (Cameron 2010: v). There is an "almost constant underlying state of uncertainty experienced by many disabled people in relation to how they are being perceived and received by others around them; something like the easily-awakened awareness of the likelihood that they are being watched and found wanting, or seen as deficient, incompetent and unfortunate." As Cal Montgomery notes, "Every few hours I run up against people who feel free to remind me that I'm their inferior and that I should conform to whatever they've decided…people like [me] are supposed to be like" (Cameron 2010: 2).

The so-called "Medical model" of disabilities makes claims somewhat like this:

> You are deficient.
> You are the problem.
> I, the professionalised servicer, am the answer.
> You are not the answer.
> Your peers are not the answer.
> The political, social and economic environment is not the answer.
> (McKnight 2005 in Cameron 2010: 7).

One reaction to this is the "supercrip" perspective that "is often lauded by non-disabled individuals, and frames individuals with disabilities as having 'overcome' their disability and are viewed as 'superhuman' because of achieving unexpected accomplishments and perhaps as having 'achieved' or 'reached' disability identity synthesis. On the contrary,…the supercrip model is resented by many people with disabilities who are simply trying to lead their lives, and the complexity of disability identity development indicates a fierce rejection of this narrative" (Forber-Pratt et al. 2017: 12).

In contrast, Cameron and others propose an *Affirmative Model* which regards "impairment as a valid human characteristic among other human characteristics. This does not involve a denial of the sometimes-painful aspects of impairment, but neither does it regard these as marks of inferiority. The affirmative model provides the basis for a self-respectful stance to be taken in the face of cultural assumptions of personal tragedy. It contextualises the experience of impairment within current discourse on diversity and establishes the rights of people with impairments to be recognized and valued as who they are" (Cameron 2010: 257).

Noting that most of the literature on disability (and all other forms of identity) is based on varieties of postmodern theory, Davis proposes to replace "the notion of postmodernism with something I want to call "dismodernism." (Davis 2013a: 265). "The watchword of dismodernism could be: Form follows dysfunction." "What dismodernism signals is a new kind of universalism and cosmopolitanism that is reacting to the localization of identity. It reflects a global view of the world. To accomplish a dismodernist view of the body, we need to consider a new ethics of the body" by creating "a new category based on the partial, incomplete subject whose realization is not autonomy and independence but dependency and interdependence. This is a very different notion from subjectivity organized around wounded identities; rather, all humans are seen as wounded. Wounds are not the result of oppression, but rather the other way around." "It is too easy to say, 'We're all disabled.' But it is possible to say that we are all disabled by injustice and oppression of various kinds. We are all nonstandard, and it is under that standard that we should be able to found the dismodernist ethic. What is universal in life, if there are universals, is the experience of the limitations of the body" (Davis 2013a: 276).

*Poems with disabilities*
By Jim Ferris

I'm sorry—this space is reserved
for poems with disabilities. I know
it's one of the best spaces in the book,
but the Poems with Disabilities Act
requires us to make all reasonable
accommodations for poems that aren't
normal. There is a nice space just
a few pages over—in fact (don't
tell anyone) I think it's better
than this one, I myself prefer it.
Actually I don't see any of those
poems right now myself, but you never know
when one might show up, so we have to keep
this space open. You can't always tell

just from looking at them, either. Sometimes
they'll look just like a regular poem
when they roll in . . . you're reading along
and suddenly everything
changes, the world tilts
a little, angle of vision
jumps, your entrails aren't
where you left them. You
remember your aunt died
of cancer at just your age
and maybe yesterday's twinge means
something after all. Your sloppy,
fragile heart beats
a little faster
and then you know.
You just know:
the poem
is right
where it
belongs (Ferris 2013).

Used with permission from the author.

## 6.4 Carceral

It is impossible to discuss all of the labels that people may choose to identify themselves or which others use to identify them. But I do want to add one group of people who I think are too pervasive and ignored to omit, and who also seem to be in the initial stages of consciously reflecting on who they are—namely, people who are or who have been incarcerated in jails and prisons. This is a very large number of people in the United States, but significant in most other parts of the world as well. The US has by far a higher percentage of its population in prison than any other nation—certainly astronomically above any other "advanced" country.

The proportion of people in jails and prisons compared to their numbers in the population is badly skewed against people of color as well: In 2017, Blacks represented 12% of the U.S. adult population but 33% of the sentenced prison population. Hispanics represented 16% of the adult population, while accounting for 23% of prison inmates. On the other hand, Whites accounted for 64% of adults but only 30% of prisoners (Pew Research 2019).

"People between ages 26 and 35 were 3.6 times more likely to have been arrested as compared to those who were at least 66 years old. About one-third of the men between the age of 26 and 35 had been arrested during their youth, 2.6 times the rate of those 66 and older." Men are far more likely to be imprisoned than women, but women have experienced an even more rapid relative increase in arrests. Among those aged 66 and older, arrests before age 26 occurred among only one in 100 participants. But among those aged 26 to 35, about one in every seven women had been arrested at least once by age 26 (Rand 2019).

As I stated before while discussing Wendell Bell and Affirmative Action Sentencing, there is little reason to assume that young nonWhite males misbehave more frequently than older White females by such orders of magnitude. Rather, it is the case that certain acts that have been criminalized are treated far more severely by the legal system than other acts, and that young, nonWhite males are far more likely to be stopped by police, questioned, apprehended, jailed, taken to court, and imprisoned for longer times than are White people of any age or gender.

Roughly two in five incarcerated people have a disability of some sort. Nearly one in four have cognitive disabilities (Sarrett 2021).

"While it is difficult to ascertain whether poverty makes someone more likely to commit a crime, data show it does make a person more susceptible to being arrested and more likely to be charged with a harsher crime and to receive a longer sentence. Adults in poverty are three times more likely to be arrested than those who aren't, and people earning less than 150% of the federal poverty level are 15 times more likely to be charged with a felony—which, by definition, carries a longer sentence—than people earning above that threshold" (Hayes and Bornhorst 2020).

Almost half of all people in federal prisons are there because of drug offenses. Twenty percent are imprisoned because of weapon possession or use. For ten percent, it is some sexual offense. Homicide and aggravated assault make up only a little over three percent of the federal prison population, and the same percent for robbery.

Reasons for imprisoning people who broke the law have varied over time. Early in America's history, incarceration was seen as a way to allow people to reflect on the error of their ways; to confess their sins, to become contrite and penitent, and then to be able to rejoin lawful society—a goal captured in the term "Penitentiary". Later, "Reform" was the key word. Law-breakers were guided through various activities that were intended to enable them to move beyond habits and behaviors that caused them to misbehave, and to acquire the habits and behaviors of law-abiders. However, for the past thirty years, the primary rationale for imprisoning people in the US has not been to enable them penitently to reflect on their crimes, or to learn law-like behaviors. The main purpose of imprisonment is punishment. Justin Smith and Aaron Kinzel write of the myriad "pains of imprisonment". They quote Campbell and Schoenfeld who state that the current situation

> is characterized by a set of ideas, including that the purpose of prison is incapacitation and retribution, that criminals are "other" and not worthy of redemption and that being labeled soft on crime is the ultimate political liability or conversely that tough-on-crime credentials are a political necessity. It contains a specific "penal field" where penal policies are shaped by politicians, law enforcement, and victims' organizations rather than by judges, social workers, criminologists, or community organizations.... And it includes penal policies with one primary goal: to incarcerate or "supervise" masses of criminal offenders for long periods of time. As a result, the...penal order is distinguished by uniquely high and racially disproportionate incarceration rates and an expansive correctional supervision apparatus (Smith and Kinzel 2020: 95).

This is called “Just Deserts”, as in giving people what they deserve. No coddling, no educating, just punishing severely: giving them their just deserts.

Of course numerous calls for reform are constantly heard, including from professionals within the system itself. Moreover, rates of imprisonment were in fact falling until very recently, and significant prison reform seemed for a while to be on track as a strongly bipartisan movement. Between 1972 and 2009, the US national prison population grew by 700%—from roughly 200,000 to over 1,500,000. Then from 2009 to 2018, it declined by nine percent to 1,400,000. However, even though crime rates declined by half from 1990, rates of incarceration remained disproportionately high. The “Slap ‘um, ‘n’ rap ‘um” practice of Just Deserts still is the prevailing philosophy, especially during the reign of Trump and Covid-19. They haven’t changed under Biden. Hatred produces hate and systemic injustice provokes hateful revenge.

White reactions to “Black Lives Matter” and “Defund the Police” seem to have killed the possibility of bipartisan, humane reform. At present, the American public seems more sharply divided on the issue than ever before. At the same time, conversations about the identity of incarcerated persons appear to be increasing as well. The consciousness of current and past unjustly-incarcerated persons may very well follow a trajectory from rigid, ascribed, and embraced wholly negative identities, to fluid, positive, prideful self-identities similar to what we have seen for gender, race, class, and disability.

Indeed “prisoner identity” now seems to be about where “disability identity” was several decades ago when disabled persons were pitied; told they were “good little soldiers” for doing their very best to “fit in” as “normal people”. Now, many take pride in themselves, saying that it is the behavior of others, and the structures and processes of their environment, that turned them into “handicapped people.” Remove the handicaps, enable their abilities, and they can and will help swell the rich diversity of humanity beyond narrow “normality”, they say. Something like that may be happening with prisoners and especially former prisoners.

> The Marshall Project has developed a policy…of ‘people-first’ language. Originally developed by people with disabilities, people-first language avoids turning one aspect of a person’s life into an all-encompassing label.” “Words like ‘inmate,’ ‘prisoner,’ ‘convict,’ ‘felon’ and ‘offender’ are like brands. They reduce human beings to their crimes and cages.

Based on surveys of what people in prison wished to be called, the Marshall Project developed a policy using terms such as “incarcerated people”, “imprisoned people”, “people in prison”, “people in jail”, “people jailed in X facility.” “formerly incarcerated people”, “John Doe, who was incarcerated at FCI Memphis…” “Jane Doe, who is serving 12 years in San Quentin State Prison…” “Held in Rikers Island Jail for three years without a trial, Kalief Browder…” “A 34-year-old detained in Los Angeles” and the like (Solomon 2021).

Most former and current incarcerated persons serve long sentences for what might be called “status offenses” that harmed no one, and should have been handled in different, more effective, and humane ways. Moreover, every day, more and more people are being exonerated after serving decades for crimes they did not commit—as they have long protested. Just what do they have to be “penitent” about? What reforming brainwashing do they need to endure? What just deserts do they deserve? The answers to these questions suggest that something like "prison pride"—wearing imprisonment as a badge of unsung courage like Black Pride, Gay Pride and the rest—might not be far behind.

We are not there yet, but the trend towards diverse, fluid, constructed identities by former "prisoners" seems clear as it does for race, gender and disability.

Indeed, if America continues to slide down the syrupy slope towards violently-competing tribal dictators, we can imagine that America also will begin to have more and more people imprisoned for their political beliefs and acts. This might especially be the fate of presidents when they are removed from office, perhaps legally via biased election procedures with judicial and vigilante support. One consequence might be even more political prisoners than there are now who are righteously outraged by their incarceration, and proudly bonded to be alumni of US prisons. They, in turn, will seek revenge when it comes their turn to rule, imprisoning (or worse) their opposites. Having spent time in prison may increasingly be a mark of pride and retribution than it might otherwise.

Justin Smith and Aaron Kinzel suggest a way forward, though they do not point to it specifically themselves. They document "the unprecedented psychological and collective harms instituted by the carceral state over the last several decades in the US: (1) the sheer numbers and racial disproportionality that constitute mass incarceration; (2) the psychological harms that result from conditions such as the absence of individual autonomy, overcrowding, solitary confinement, and threats of violence; and (3) the post-imprisonment collateral consequences of political disenfranchisement, as well as restrictions relating to employment, health care and housing that accumulate to form a state of "carceral citizenship" (Smith and Kinzel 2020: 94).

At the same time, Smith and Kinzel highlight that carceral citizenship also "allows some (perverse) benefits, including 'access to goods and services reserved for formerly incarcerated people and the symbolic benefits of public regard for those who have made good'. These symbolic benefits are apparent when individuals use their first-hand experience of incarceration and apply it toward justice activism. While their analysis rightfully focuses on how 'carceral citizenship' relegates individuals to an alternate legal reality, the observation that carceral expansion has produced an alternative citizenship category also suggests the potential socio-political power of some within this collectivity."

Finally, Smith and Kinzel's concept of carceral citizenship "relies heavily on convict criminology to illustrate the importance of formerly incarcerated voices in the transformation of the carceral system…; on the framework of restorative justice to argue for the inclusion of those who have experienced (and suffered from) incarceration within collective actions and organizations working toward transformation of the carceral system…; and on constitutive criminology, which demonstrates that recognition of the lived experience of marginalized groups moves us toward 'replacement discourses' in efforts to reconstruct discourse of 'crime' and 'law'. We use these various criminological perspectives to conceptualize how 'carceral citizenship' contains advantages and not only disadvantages" (Smith and Kinzel 2020: 94).

As is the case for most identity studies, evidence here comes largely from narratives, from stories that people tell about themselves and their experiences with ascribed as well as chosen identities—the obstacles, the joys, the rage, the strengthening of character and resolve as new, forbidden, almost unimaginable identities arise to refute the stories that those in power forced upon them for years.

Michael Torres' poem, "My brother is asking for stamps", tells a story of incarceration from the outside, about his imprisoned brother at first optimistically looking forward to resuming normal life. He is sorry he couldn't attend a wedding, "He says it's not so bad in here,/says he's not getting institutionalized,/won't get institutionalized, not like/the others". But

> It's May. It's March. It's May. It's October.
> Happy Halloween, Brother. He's asking,
> again, for postage stamps, telling me
> he might be programmed, sure, but
> who isn't?

And on through the months and years until

> My brother
> says he'll write when he can, he knows
> I'm busy. Everyone's busy. It's August.
> It's August, and he's looking for stamps (Torres 2020).

## 6.5 Refugees

Of all the identities that I have discussed so far, none is more generically concerned with "who am I?" than refugees. We saw many pages ago when we first discussed "identity" that "who am I?" is often considered to be the paradigmatic question that identity seeks to answer. The very fact of being a refugee from someplace to some other place is disruptive, and the relationship of one's identity in a newfoundland compared to what it was in the homeland is initially traumatic and typically perpetually problematic. Unlike immigrants who move voluntarily from one place in order to relocate in a new "forever home", refugees are, initially, asylum seekers who were either forced out of what they considered to be their home or fled for their lives seeking to be let in and protected by other governments or caregivers. Not all asylum seekers are able to become refugees—they may not have the marks necessary to be designated a refugee, and/or be refused the status by agents of the governments whose protection they seek—but all refugees are expected both to be seeking asylum from specific dangers and to meet the criteria of the receiving country in order to be admitted.

Article 1 of the United Nations 1951 Convention Relating to the Status of Refugees defines a refugee very explicity as a person who,

> owing to a well-founded fear of being persecuted for reasons of race, religion, nationality, membership of a particular social group or political opinion, is outside the country of his nationality and is unable or, owing to such fear, is unwilling to avail himself of the protection of that country; or who, not having a nationality and being outside the country of his former habitual residence as a result of such events, is unable, or, owing to such fear, is unwilling to return to it. (United Nations 1951).

Refugees are unusual in that their identities and the services that they have a right to receive are determined by international law, and are not primarily the result of diverse cultural, historical, or personal preferences and controversies as many identities are. However, it is important to realize that contemporary international law, though supposedly of global jurisdiction, is primarily a western concept, conceived in western terms and experiences to solve primarily western problems: "The flight of so many peoples in the early part of the 20th century, Greeks, Turks, Armenians, and victims of the First World War, together with a rather restrictive approach to immigration, forced the hand of states…to support a framework aimed at assisting refugees. The League of Nations, founded in 1919, assumed responsibility for fulfilling the aspiration of an international refugee regime. … From a legal perspective, one of the notable features in the development of international refugee law is the move from a group based analysis, in the period prior to the Second World War, to an individualised approach incorporated post-War" with the creation of the International Refugee Organization by the United Nations. "The focus on the individual…was also reflected by the Preamble to the United Nations Charter 1945 in its emphasis on the 'dignity and worth of the human person', and fully endorsed by the Universal Declaration of Human Rights 1948." "The individualised approach to refugee status is undoubtedly a key component of post-1951 legal identity; in the main, the refugee was to be assessed on his or her own merits, not as part of a wider ethnic group." The "legal identity of the refugee has not remained static; rather, it has evolved, and continues to evolve, as a direct consequence of the movement of peoples across borders, whether individually or *en masse*, and of the changing way in which such movement is perceived by host states" (Stevens 2014: 3, 7).

In Europe, in the years before the Second World War, many Jews left Germany and allied countries as Nazi policies hardened and deepened. Many sought refuge in neighboring countries in Europe only to be forced to flee elsewhere, and many who did not or were not able to leave were killed. Few countries, including the United States welcomed the refugees, placing many obstacles in their way. The US Holocaus museum states that historians "estimate that approximately six million Jews were murdered during the Holocaust, including approximately 2.5 in killing centers, two million in mass shooting operations, and more than 800,000 in ghettos" (United States Holocaust Memorial Museum 2022).

However, immediately after World War II, a disaster of gigantic proportions happened elsewhere. "In August, 1947, when, after three hundred years in India, the British finally left, the subcontinent was partitioned into two independent nation states: Hindu-majority India and Muslim-majority Pakistan. Immediately, there began one of the greatest migrations in human history, as millions of Muslims trekked to West and East Pakistan (the latter now known as Bangladesh) while millions of

Hindus and Sikhs headed in the opposite direction. Many hundreds of thousands never made it. Across the Indian subcontinent, communities that had coexisted for almost a millennium attacked each other in a terrifying outbreak of sectarian violence, with Hindus and Sikhs on one side and Muslims on the other—a mutual genocide as unexpected as it was unprecedented. In Punjab and Bengal—...the carnage was especially intense, with massacres, arson, forced conversions, mass abductions, and savage sexual violence. ... By 1948, as the great migration drew to a close, more than fifteen million people had been uprooted, and between one and two million were dead. ... the British Army was able to march out of the country with barely a shot fired and only seven casualties" (Dalrymple 2015).

Wherever existing nations are partitioned—as they were in Africa before, during, and after the period of western colonization or where nations were divided between capitalist and communist regimes—refugee flows occured in great or lesser degrees. Uditi Sen provides an interesting example of what might be called "refugee power" during the Partition of India. "Within the popular memory of the partition of India, the division of Bengal continues to evoke themes of political rupture, social tragedy and nostalgia. The refugees, or more broadly speaking, Hindu migrants from East Bengal, are often the central agents of such narratives. This article explores how the scholarship on East Bengali refugees portrays them either as hapless and passive victims of the regime of rehabilitation, or eulogises them as heroic protagonists who successfully battle overwhelming adversity to wrest resettlement from a reluctant state. This split image of the Bengali refugee, as victim/victor, obscures the complex nature of refugee agency." "Scattered throughout the reminiscences of the squatters are anecdotes of everyday resistance, negotiation and accommodation, which together provide a far richer and complex understanding of refugee power" (Sen 2013: 14).

While the definition of a refugee in international law is relatively clear, it is understood and administrated differently depending in part on where the refugees are from and why they left, as well as the culture and political interests of the would-be host nation. Stevens' paper that we referenced above is focused primarily on the differences between the way European nations generally interpreted and actualized the laws legalistically with the way some Arab states acted: "The history of the Middle East, however defined, is one of conflict, enormous upheaval and mass movement of peoples, and a basic understanding of the historical roots of displacement is crucial when exploring centuries of forced migration in the region." "While the (contentious) history of Armenian massacres in the late 19th century and early 20th century is well known, in which it is claimed that between 1.5 and 2 million people were uprooted, and anywhere between 600,000 and 1.2 million died, widespread ethnic or religious cleansing or the so-called 'unmixing of peoples' undertaken in the same period is less familiar." "[W]e find numerous accounts of the forced migration of hundreds of thousands of people." "It is claimed that 27% of the Muslim population of Ottoman Europe were forced to move as a result of the Balkan Wars, while 500,000 perished on the route from the Caucasus to Anatolia. And this is only a snap-shot of the many migrations and re-migrations that occurred in the region over a relatively short period of history" Stevens 2014). Given the incessant wars and coups in the region subsequently, refugee flows out of one Arab country into another as well as into other parts of Africa, Europe, and elsewhere have been continuous.

Among the most well-known and long-lasting refugees are the Palestinians. "People may feel an intense subjective affiliation with a national identification, particularly in a case like that of the Palestinians, deprived of their homeland by colonialist negation. But even though the duration of their exile increases Palestinian refugees' desire for Palestine, reinforcing their sense of shared identification, yet it simultaneously deepens colorings of 'self-identification' though different locations, class positions; politics and ideologies. This poses the likelihood of a growing gap between a shared 'Palestinianness' and differing class, regional, political or individual interests."

"Some reject the term refugee as humiliating, but others affirm it as signifying that they belong in Palestine and not in Lebanon. From a recent recording in Bourj al-Barajneh: "Q: The word refugee, do you accept it as a part of your identity, or do you refuse it? A: Of course (I accept). The word refugee means that I'm Palestinian, and it specifies from which area. We must cling to it". However another speaker said: "No, of course I refuse it. It means that others are higher than me. We're on the ground, they are in the sky" (Sayegh 2012: 13, 19). "The refugees are sick of being the quintessential refugee question for all of these years. They just want to live and to be allowed to live." Laila, a 35-year-old Palestinian refugee in the West Bank (Beckerle 2012: 13).

Turning to Stevens again, she says "A fascinating example is provided by the Iraqis who fled the first and second Gulf Wars in 1991 and 2003…." "Interestingly, certainly in the early days of the migration, Iraqis preferred to be seen as guests, since they too associated the 'refugee' nomenclature with a particular form of refugeehood—that of Palestinian hardship and victimhood. The idea of the refugee as guest has deep-seated cultural and religious roots." "The Qur'an requires that refugees and migrants be welcomed and treated well, and should not be refused admission, rejected at the borders, or sent back to the country of origin." "The obligation to grant asylum applies to Muslims and non-Muslims alike, to rich and to poor. It is non-discriminating. In Islam, both the State and the individual can grant asylum." "The Qur'anic principle of asylum is not only a religious tenet but has been legalised through Shari'ah Law." (Stevens 2014: 20, 22).

While the identity of being a refugee may seem clear cut in international law though flexible in administration, asylum seekers and refugees literally come from everywhere in the world and seek refuge anywhere else. Similarly, refugees from many very different parts of the world, fleeing from different specific regions, with different languages, cultures, talents, and dreams may all end up in exactly the same spot. If so, they need to negotiate not only the relationship between their cultural expectations and those of the host culture (which may be monocultural or richly culturally diverse) as well as with those of all the other persons with whom they must interact. Differences in skin color, accent, clothing, food, class, caste, education, age, gender, sexual orientation, disability, and previous condition of servitude within each of the different interacting people results in a far greater degree of identity confusion—or richness—than that of any of the other identity groups or individuals considered. However, "In contrast to the mostly 'risk'-focused literature on refugees, the extant literature on biculturalism and adaptation points to positive outcomes of living in 'two worlds,' especially what is known as the 'immigration advantage' or 'biculturalism paradox'" (Hayes and Endale 2018: 284).

As a consequence there is something that can be called the "emergent newcomer identity theme": "You create your own culture here, because you do not belong there anymore and you do not belong here. So, you create your own world. Typically, I tend to have friends that have a similar background as myself. My friends live here, but they are from another place. We talk about there and we live here, and we do not belong anywhere, except the place you have invented, your own small world." However, they quote Nedim Karakayali as commenting that "The real problem with the two-worlds thesis is not its argument that immigrant children feel caught between two worlds, but its failure to note that this experience follows from the condition of living in a world where most people believe that there are only two worlds" whereas there are vastly more (Hayes and Endale 2018: 288, 289).

"This tension highlights the importance that host communities, policy, and culture play in the process of identity formation. For example, studies conducted in the United States have a different context than those in Canada, especially as it relates to cultural practices and policies about 'assimilation' versus 'acculturation'". "...Canadian legal processes and cultural attitudes tend toward acculturation and 'mosaic approaches' to multi-culturalism, in contrast to the U.S. approach of assimilation and a 'melting pot'" (Hayes and Endale 2018: 285).

Shannon Daniels states that "Although some views of identity assume that traits could be assigned to cultural groups and one's identity remains consistent across time and space, more nuanced views of identity reject these notions and suggest that identity is fluid, multifaceted, and informed by one's view of oneself and the social construction of identity in relation with others and the world." Each person's identity includes a multitude of selves "that each person continues to deepen, expand upon, and understand his or her identity across the life span." ..."resettled refugees often experience discourses of essentialization that unfairly and inaccurately homogenize groups of individuals."

"[A]gency is central to identity and agency involves not only a response to one's past and current worlds, but also a vision for one's future". Her work with refugee youth focused on "the visions youth have for their futures and how they see themselves in terms of being able to make their visions a reality from which three major themes emerged. ... First, the youth were particularly interested in improving the lives of those around them. Second, the youth showed a deep awareness of social problems and a willingness to transform society. Finally, the youth clearly held strong beliefs in their capabilities and were envisioning rich futures for their lives. Although these three themes are overlapping and intertwining in their nature, they show how the teenagers demonstrated agency and shared their identity in how they talked and wrote about their futures." "All of the teenagers recognized that accomplishing their dreams would be difficult while also showing confidence in their own abilities to persevere and do what is needed to enact their visions of the future" (Daniel 2019: 75, 79).

Similarly, Laura Moran worked with refugee children in Australia. "[T]hree siblings of Karen descent discussed how best to characterize their own sense of national and cultural belonging. The eldest, Catalina, explained, 'I just call myself Karen. I am not white people.' Her younger brother, Thakin, answered, 'I'm Australian. I wanna be Aussie!' And Jessica, the youngest, offered, 'I have both. We like both foods now, we like Aussie food too, don't we?' They went back and forth in this way for some time. Each sibling was challenging one another's perspective on what establishes belonging—what elements of a person's experience or background constitute a sense of identity. … It was certainly apparent that a range of cultural, ethnic, and racial influences coalesced in the formulations of these young people's sense of themselves. However, they oscillated in their expressions of identity between emphasizing their ability to pick and choose from such influences—to "freestyle," as some of them described it—and downplaying this flexibility in favor of presenting their racial and ethnic identity as fixed and binding. "By representing themselves in racialized ways and alternately 'inhabiting' or 'vacating' their racial and ethnic backgrounds, young people enabled a multiplicity of options for how and to whom they might assert a sense of belonging." "[T]heir emphasis on sameness and difference through hybridized and essentialized representations was not static. As they engaged with tensions of belonging, their essentialized and hybridized representations often merged, overlapped, and occasionally contradicted one another. And in their use of a range of cultural preferences in the presentation of themselves as a cohesive, essentialized group, these young people were at the same time engaging in hybridizing practices" (Moran 2020: 141, 143).

The writing and activism of Hannah Arendt after World War II has had tremendous influence in many areas, including refugees. "Refugees driven from country to country represent the vanguard of their people—if they keep their identity". According to Cindy Horst and Odin Lysaker, "Arendt is preoccupied not only with dark times, but also with what she terms 'miracles' and 'hope' for the future, which is stated in *The Human Condition*. She is concerned not only with the breakdown of citizenship rights and the loss of home, occupation, language and community, but also with what she takes to be humans' freedom in terms of natality." "By natality, Arendt means several things, such as being born into a shared world and, through becoming newly born, introducing something entirely unique into that world with respect to an individual's diverse characteristics (i.e. the human condition of plurality). The human condition of natality, to Arendt, is what she terms as 'new beginnings' or 'beginning anew', which involves a freedom to act, but also to raise questions about customs or traditions and a capacity to introduce innovative and unpredicted ways of acting and interacting in the world." "The Arendtian definition is promising for understanding the figure of the refugee as pariah as well as vanguard…. New beginnings involve a freedom to act, to introduce innovative and unpredicted ways of being and acting in the world. This freedom is expressed clearly in the actions that refugees take—to redress the situation of other refugees or marginalized groups, for example—that show a deep sense of responsibility as well as an inability to do nothing. This freedom to act and interact offers a clear hope for a future of new beginnings and miracles" (Horst and Lysaker 2019: 70, 71, 82).

Lawrence-Minh Bui Davis and Ocean Vuong are the co-founders of the amorphous Center for Refugee Poetics. Davis quotes from "Self-help for fellow refugees" by Li-Young Lee. Lee's poem describes a son watching his father being hauled away by "armed men" whereupon his mother buries the boy's face in her skirt to shelter him from the wrenching scene. Davis ends a fragment of the poem:

> And I bet you can't say
> what language your father spoke
> when he shouted to your mother
> from the back of the truck, "let the boy see!" (Davis 2022).

## 6.6 Environmental Refugees

There is no such thing as "environmental refugees" even though they exist and have pleaded for aid and relief before various national and international tribunals with no success. Neither the term, "environmental refugee", nor anything like it is included in the international law documents that define who refugees are and what rights they have. Therefore they do not exist in the formal, judiciable way that refugees do. The term does not appear to have been written into the law of any nation. The breadth of the term even informally is highly contestable. On its face, the term "environment" combined with "refugees" seems to suggest people (in groups or as individuals?) who are forced to leave their home and seek settlement elsewhere because of a wide range of possible, persistent, existing or future environmental factors—from drought, famine, floods, fires, and hurricanes to climate change that makes it impossible to plan and act on the basis of past climate patterns, and sea level rise and intrusion that is devouring costal zones in most parts of the world as well being on the verge of swallowing entire island nations in the Pacific, Indian, and other oceans. "There is no consensus today on a definition of the term 'environmental refugees' since 1985 when it officially appeared. Several descriptions such as ecological refugees, environmental refugees, climate refugees, eco-refugees, climate évacué, environmental migrants, displaced persons due to a natural disaster, environmentally displaced persons" exist (Sahinkuye 2019).

As we have seen, The United Nations 1951 Convention Relating to the Status of Refugees defines what a refugee is. "There are four main components of this definition. The first is that the person must be outside their country of nationality or former habitual residence. Second, the person must fear persecution. Third, the fear of persecution must be for reasons of 'race', nationality, religion, membership of a particular social group or political opinion. Fourth, the fear must be well founded. This definition and its components do not include individuals who flee their homeland or habitual residence because of environmental change" (McNamara 2007).

Who the phrase covers and what are considered the causes for people to flee their land have changed considerably over time. "Much of the [early] writing on the topic introduced neo-Malthusian concerns, linking population growth with social and economic crisis, and with out-migration" (Morrissey 2012). Since then the term has taken on many meanings. At the present time it often refers to migrants who are already suffering from, or believe they are about to be dislocated by, the effects of global warming and climate change, including significant sea level rise. This is especially the case for the small island nations of Kiribati and Tuvalu in the south Pacific. These are fragile atolls where the highest point of ground until recently was only about two meters above mean sea level, often experiencing damage from waves and flooding during certain high tides and especially hurricanes. Though independent nations by international law, these two have a special relationship with New Zealand. Many of their citizens work in New Zealand or frequently travel there. They have sought relief, so far unsuccessfully, from courts and tribunals in New Zealand. They and other Pacific Islanders have pleaded their case before the United Nations. Representatives from the Maldives, in the Indian ocean, have also actively sought anticipation action from the UN. Newtok, Alaska, has been losing ground to the sea at such a dangerous rate that land was acquired for a new townsite called Mertarvik on Nelson Island about 14 km away. "The goal is to complete relocation by 2023" (Newtok 2021).

This is not just an international phenomenon. Americans are already on the move within the US because of the impacts of climate change now, including wildfires, floods, droughts, and searing heat. Like all refugees, they are facing rejection and resentment as well as welcome in their new homes. Jesse Keenan of Tulane University's School of Architecture estimates "that 50 million Americans could eventually move within the country to regions such as New England or the Upper Midwest in search of a haven from severe climate impacts. He predicted that migration driven by increasingly uninhabitable coastal areas is likely to happen sooner rather than later…" (Hurdle 2022).

Much of the literature on environmental refugees, whether pro or con the designation, points out that there is no clear perpetrator for victims to blame, as is required in order to claim to be a refugee. Elizabeth Keyes points out that "The Convention definition of a refugee requires not just that people face severe harm, but that the harm is happening *because* of a protected characteristic: race, religion, nationality, political opinion, or membership in a particular social group. This nexus between the harm and the protected characteristic is where a casual use of the term "environmental refugee" reveals its imprecision." "But who is the agent persecuting the people of Kiribati? Under the Refugee Convention, it needs to be either the government itself, or a private entity the government is unable or unwilling to control" (Keyes 2019: 4, 5).

Francois Gemenne adds that "Even if humans have indeed replaced natural drivers of changes as the principal agents of changes on this planet, most humans are actually the victims of these changes, and not their agents." "In our attempt to stress the agency of the migrants, we had forgotten the responsibility that we had towards them, because we humans have become the main agents of transformation of the Earth. And the result of this transformation has been to make their places on the Earth increasingly uninhabitable for a growing number of people." "Industrialised nations have thus little incentive to act; our agency is undone by our self-interest" (Gemenne 2015: 1, 2).

Keyes, Morrissey and others show that there are ways the existing law on refugees might be interpreted to include environmental refugees. For example, Keyes states, "Government actions and inactions on climate often affect different sub-groups of the country differently and are classic examples of discrimination; such discrimination might rise to the level of persecution if it means members of those sub-groups are reasonably fearful for their future existence." "[C]limate change will likely give rise to new political groups and dynamics and, therefore, to political opinions opposed to the direct and indirect government policies and actions on climate-change.... [A] party member motivated by a climate-change issue, who fears persecutory retaliation from the government *because of* their political opinions, presents a straightforward application of the Refugee Convention." "[W]omen may endure special kinds of harms, distinct from the general population." The Convention states that it is necessary "to prove that the feared harm would happen *because of* that protected characteristic." "It is beyond reasonable dispute that structural racism explains why African-Americans were more vulnerable to Hurricane Katrina's impacts than any other group in New Orleans" (Keyes 2019: 8, 12–15).

Morrissey in his review of the contentious positions about the existence and proper status of people who flee their homes because of existing or impending environmental destruction concludes that in spite of all the disagreements, "The 'debate on environmental refugees' is really a debate about *how a relationship is represented*, and less about the *nature of the relationship itself*." "There is virtually no debate about whether environmental change, or stress, impacts on decisions about human mobility. Such a position is accepted, even by the fiercest critics of the notion of an 'environmental refugee'" (Morrissey 2012: 40, 45).

That may be so among scholars primarily focusing on environmental refugees, but there are many other people who deny there are any environmental reasons for migration: climate change is a hoax, an economic scam, a communist plot to stop capitalist economic growth, to which others also add that it is a clever way to permit undesirable immigration.

As we saw, refugees seeking asylum from the horrors of Nazi Germany and, later, from Communism were typically lauded and greeted with open arms in the US and elsewhere. But with Fall of the Wall in 1990, the "flood" of refugees from the Balkan countries engulfed in civil war, and the flight of Syrians and others out of the Middle East and then various killing fields in Africa, anti-immigrant sentiment began to grow and is still growing. Donald Trump very successfully demonized asylum seekers from South and Central America as thieves and rapists who would become dependent on the honest labor of real Americans for their food, board, education, and health care—when they were not actively stealing real jobs from real Americans and working, poorly, for less pay and in squalid situations.

At this moment of writing, some of the world is outraged by the military attack of Ukraine by Russia and the resulting flow of refugees from Ukraine to far-off as well as neighboring nations. Some observers have pointed out how different America's prompt welcoming of these White brothers and sisters is compared to the walls of rejection built along America's southern borders and the near exclusion of refugees of color from the Middle East and Africa to equally-horrifying barbaric slaughter.

Most refugees after World War Two until the collapse of communism in 1990 were White. Most since then are not. Strong sentiment against environmental refugees enables climate change deniers and racist anti-immigrants to join forces, if they are not basically the same people to begin with.

On the other hand as more and more people in so-called "developed nations" personally experience the tragedies of climate change, the distinction between who caused the disruption and who suffers from it that marks the debate so far may change. There is no doubt that the rich and powerful will use their resources to buffer themselves from the worst consequences of climate change for as long as they can, but if the realities of the Anthropocene Epoch are as widespread, rapid, and disruptive as I think they will be, there may be no place on or off Earth where anyone can hide. At the present time, the winning position between welcoming and denying environmental refugees may all be on one side. It seems to me we may all be environmental refugees running in circles looking for safe haven, unless we respond more positively and powerfully to climate change than anyone is doing or seriously proposing now, and/or until we all find other niches on other planets that are more hospitable to humans with their current biological/environmental limitations as they are now.

Of course the entire controversy about immigrants, asylum seekers, refugees and environmental refugees is based on the pathology of past- and revenge-oriented identity sentiments and politics that I seek to have us to transcend. Why do we unthinkingly perpetuate—worship—the form and powers of nation-states as we do? Surely there are many other forms of governance beyond the nation-state—as well as beyond democracy or any other current or historical system that is more appropriate for the challenges and opportunities of the Anthropocene Epoch.

As a political scientist, I have taught graduate and undergraduate courses on new governance design for fifty years. Among other things, I encourage my students to ignore the historical and current philosophies, ideologies, structures and processes of governance and re-think them from all from the basic assumptions and foundations on up, starting with the basic unit of analysis today. The unit of analysis for all current governance is the obsolete European sovereign nation-state system that was created several hundred years ago to solve a political problem in Europe. It is an unnatural, awkward invention that causes a great deal of harm, internally and externally. It does not naturally fit many communities now. China does not consider itself to be merely one nation among others. The fundamental unit for Islam is the Ummah (أمة)—the community of believers worldwide. Both China and Islam have learned how to fit into the old European system for a while, but many people, such as myself, feel affinity both to more local communities (Honolulu) and globally to the Earth and all life on it, and not to the United States of America. Every time I must pass through border control I feel demeaned and insulted by agents of the US government as well as by those of all other governments. Why not consider something other than the nation-state as the basis of governance design? Why not make the Unit of Analysis—the individual person, family, identity group (ethnic, religious, linguistic, occupational, virtual, or other), local physical community, bioregion, Earth, inner solar system, solar system, Milky Way Galaxy, or intergalactic cosmos? Or why not think smaller and more intimately—DNA, cells, microbes, fungi, and viruses who rule inside each of us? Or smaller still—molecules, atoms, and quarks?

Among others, Joseph Carens does not ask us to think that extragantly but made a strong case several years ago for open borders (Carens 1987). He has elaborated on his views in several fora. His fundmental position, with which I agree in principle, is this:

> In principle, borders should generally be open and people should normally be free to leave their country of origin and settle wherever they choose. In many ways, citizenship in Western democracies is the modern equivalent of feudal class privilege—an inherited status that greatly enhances one's life chances. To be born a citizen of a rich state in Europe or North America is like being born into the nobility (even though many of us belong to the lesser nobility). To be born a citizen of a poor country in Asia or Africa is like being born into the peasantry in the Middle Ages.... Like feudal birthright privileges, contemporary social arrangements not only grant great advantages on the basis of birth but also entrench these advantages by legally restricting mobility, making it extremely difficult for those born into a socially disadvantaged position to overcome that disadvantage, no matter how talented they are or how hard they work. Like feudal practices, these contemporary social arrangements are hard to justify when one thinks about them closely (Carens 2015).

His position can be made even stronger, I believe, for environmental refugees. Climate change and its consequences are in large part caused by human action, specifically by actions within the so-called developed and developing nations. It was the processes of "development" that caused the Earth's climate to change—that ended the temperate and relatively stable climates of the Holocene Epoch and thrust all of us into the profound uncertainties of Anthropocene Epoch. Development designates the processes that enabled some people in some nations to become extremely rich while many people in the rich nations, and many more people in many other nations, were emmiserated—indeed, forcibly "de-developed" and required to live marginally in marginal lands that now are the first to suffer from the initial impacts of climate change. There is no doubt these pioneers should be the first also to leave their homes and seek refuge in places that caused them to flee but have not yet had to suffer the consequences of their behavior.

## References

Bauman, H-Dirksen L., and Joseph J. Murray. 2013. Deaf studies in the 21st century: 'Deaf-gain' and the future of human diversity. In *The disability studies reader*, ed. Lennard J. Davis. New York: Routledge.

Beckerle, Kristine. 2012. Pity versus rights' recognition: Rejection of the victim label by Palestinian refugees. In *Palestinian refugees: Different generations, but one identity*, The Forced Migration and Refugee Unit. Birzeit-Palestine: The Ibrahim Abu-Lughod Institute of International Studies, Birzeit University.

Calafell, Bernadette Marie, and Thomas K. Nakayama. 2016. Queer theory. In *The international encyclopedia of communication theory and philosophy*, ed. Klaus Bruhn Jensen and Robert T. Craig. New York: Wiley.

Cameron, Colin. 2010. Does anybody like being disabled? Doctoral dissertation, Queen Margaret University.

Carens, Joseph H. 1987. Aliens and citizens. The case for open orders. *The Review of Politics* 49 (2).

Carens, Joseph H. 2015. The case for open borders. https://www.opendemocracy.net/en/beyond-trafficking-and-slavery/case-for-open-borders/.

Dalrymple, William. 2015. The great divide: The violent legacy of Indian Partition. *The New Yorker*, June 29. https://www.newyorker.com/magazine/2015/06/29/the-great-divide-books-dalrymple.

Daniel, Shannon. 2019. Writing our identities for successful endeavors: Resettled refugee youth look to the future. *Journal of Research in Childhood Education* 33 (1).

Davis, Lawrence-Minh Bui. 2022. On refugee poetics and exophony. *Poetry* 2022, April.

Davis, Lennard J. 2013a. The end of identity politics: On disability as an unstable category. In *The disability studies reader*, ed. Lennard J. Davis. New York: Routledge.

Davis, Lennard J. 2013b. Introduction: Normality, power, and culture. In *The disability studies reader*, ed. Lennard J. Davis. New York: Routledge.

Ferris, Jim. 2013. Poems with disabilities. In *The disability studies reader*, ed. Lennard J. Davis. Routledge.

Fink, Jennifer Natalya. 2022. We should claim our disabled ancestors with pride. *New York Times*, February 27.

Forber-Pratt, Anjali, Carlyn Mueller, and Dominique Lyew. 2017. Disability identity development: A systematic review of the literature. *Rehabilitation Psychology*, April.

Gemenne, Francois. 2015. One good reason to speak of 'climate refugees'. *Forced Migration Review* 49 (May).

Halperin, David. 2003. The normalization of queer theory. *Journal of Homosexuality* 45 (2–4).

Hayes, Sherrill W., and Etsegenet Endale. 2018. "Sometimes my mind, it has to analyze two things": Identity development and adaptation for refugee and newcomer adolescents. *Peace and Conflict: Journal of Peace Psychology* 4 (3).

Hayes, Tara O'Neill, and Margaret Barnhorst. 2020. Incarceration and poverty in the United States. https://www.americanactionforum.org/research/incarceration-and-poverty-in-the-united-states/#ixzz73r0Qx9e4.

Horst, Cindy, and Odin Lysaker. 2019. Miracles in dark times: Hannah Arendt and refugees as 'Vanguard'. *Journal of Refugee Studies* 34 (1).

Hurdle, Jon. 2022. As climate fears mount, some are relocating within the US. *Wired*, April 9.

Johnson, Jenny. 2013. James River. In *Troubling the line. Trans and genderqueer poetry and poetics*, eds. T.C. Tolbert and Trace Peterson. Callicoon, New York: Nightboat Books.

Keyes, Elizabeth. 2019. Environmental refugees? Rethinking what's in a name. *ScholarWorks@University of Baltimore School of Law*.

Kolker, Zoe M., Philip C. Taylor, and M. Paz Galupo. 2020. "As a sort of blanket term": Qualitative analysis of queer sexual identity marking. *Sexuality & Culture* 24.

McCann, Hannah, and Whitney Monaghan. 2020. *Queer theory now*. Red Globe Press.

McNamara, Karen Elizabeth. 2007. Conceptualizing discourses on environmental refugees at the United Nations. *Population and Environment* 29.

Morrissey, James. 2012. Rethinking the 'debate on environmental refugees': From 'maximilists and minimalists' to 'proponents and critics'. *Journal of Political Ecology* 19.

Moran, Laura. 2020. *Belonging and becoming in a multicultural world: Refugee youth and the pursuit of identity*. Rutgers University Press.

Newtok. 2021. https://en.wikipedia.org/wiki/Newtok,_Alaska.

Pew Research. 2019. https://www.pewresearch.org/fact-tank/2019/04/30/shrinking-gap-between-number-of-blacks-and-whites-in-prison/.

Rand. 2019. Younger Americans much more likely to have been arrested than previous generations; increase is largest among whites and women. https://www.rand.org/news/press/2019/02/25.html.

Ruberg, Bonnie, and Spencer Ruelo. 2020. Data for queer lives: How LGBTQ gender and sexuality identities challenge norms of demographics. *Big Data & Society*, January–June.

Sahinkuye, Mathias. 2019. A theoretical framework for the protection of environmental refugees in international law. *The Transnational Human Rights Review* 6.

Sarrett, Jennifer. 2021. US prisons hold more than 550,000 people with intellectual disabilities—They face exploitation, harsh treatment. https://theconversation.com/us-prisons-hold-more-than-550-000-people-with-intellectual-disabilities-they-face-exploitation-harsh-treatment-158407. May 7 2021.

Sayegh, Rosemary. 2012. Palestinian refugee identity/ies: Generation, region, class. In *Palestinian refugees: Different generations, but one identity*, The Forced Migration and Refugee Unit. Birzeit-Palestine: The Ibrahim Abu-Lughod Institute of International Studies, Birzeit University.

Sen, Udidi. 2013. *The myths refugees live by: Memory and history in the making of Bengali refugee identity*. Cambridge: Cambridge University Press.

Smith, Justin, and Aaron Kinzel. 2020. Carceral citizenship as strength: Formerly incarcerated activists, civic engagement and criminal justice transformation. *Critical Criminology* 29: 93–110.

Solomon, Akiba. 2021. What words we use—and avoid—when covering people and incarceration. The Marshall Project 2021. https://www.themarshallproject.org/2021/04/12/what-words-we-use-and-avoid-when-covering-people-and-incarceration.

Stevens, Dallal. 2014. Shifting conceptions of refugee identity and protection: European and Middle Eastern approaches. Legal Studies Research Paper No. 2014-10, Warwick School of Law, January.

Tang, Terry, and Mike Schneider. 2022. Census' Asian overcount masks nuances, experts say. *Honolulu Star-Advertiser*, April 11.

Torres, Michael. 2020. My brother is asking for stamps. *Poetry*, February.

United Nations. 1951. *Convention relating to the status of refugees*. Geneva: UNHCR.

United States Holocaust Memorial Museum. 2022. https://www.ushmm.org/teach/fundamentals/holocaust-questions.

# Chapter 7
# Destination Identities

**Abstract** Creating futures-oriented identities and pathways, called destination identities. The importance of personal and societal images of the futures. The role of science fact and fiction in both inspiring and thwarting—especially in thwarting and distorting—compelling, actionable images of preferred futures for individuals and communities.

**Keywords** *Brave New World* · Destination identities · Futures studies · Games · Images · Identity · International Space University · Life stage ceremonies · Metaverse · Science fiction · Space fiction · Stories · Universities · Virtual reality

In "Destination Identity: Futures Images as Social Identity", Mohsen Taheri Demneh and Dennis Ray Morgan point out that "Very little research has been carried out on the role images of the future play in shaping identity". "A cursory look at the kind of research carried out in the field of identity reveals that the main emphasis has been placed on similarities acquired from the past. Until recently, this fundamental assumption remained unquestioned…." They add that it is ironic indeed that "people have absolutely no choice in the selection of an identity that they, nonetheless, often spend their lives defending."

However, Demneh and Morgan demonstrate that identities also originate from "images of the future, which are fluid and flexible; hence they are malleable, and once people become aware of them, they can consciously participate in social critique and reconstruction of these images of the future and achieve a new identity….Destination identity includes identifying those aspects of images of the future that can help build new identities, to create and expand on shared public images on the basis of shared values and expectations, while focusing on redefining identities based on shared images of the future" (Demneh and Morgan 2018: 56, 58, 60).

Any good futurist will tell you that the subject matter of the academic and consulting discipline called "futures studies" is not "the future". Absent time travel, or something similar, "The Future" does not exist as a thing to be studied in any sensible use of the term. What futurist study empirically are "images" of the future—hopes, fears, beliefs about "the future"—which are numerous, and found in articles, books, poems, songs, stories, prayers, movies, videos, laws, and actions. Some images



J. Dator, *Beyond Identities: Human Becomings in Weirding Worlds*,
Anticipation Science 7, https://doi.org/10.1007/978-3-031-11732-9_7

express deep myths from the origins of certain cultures but not of others. Some images are new and fleeting.

But the one arena of time and space each of us, individually and collectively, appears to have some agency and responsibility is not the past, or even the present, but the future. Certain images of the future provoke certain actions and other images provoke other actions. The impact of the actions then may modify old images or create new ones. Studying these images and their consequences is the heart and soul of futures studies. Fred Polak, a pioneering Dutch futurist, is credited with pointing out the centrality of images of the future to the fate of each culture. Certain images are positive and compelling, others are negative and discouraging. Cultures need positive images to thrive (Polak 1961).

So do individuals. Each individual's images of the futures are strongly influenced by those of their culture over all, as well as by images held and taught by their family, friends, school, religion, ethnicity, gender, class.... Each individual's own experiences in life shape their images of the future as well. Seldom are images specifically taught, but they are difficult to escape since they supersaturate the environments in which each of us live.

Mohsen Taheri Demneh and Zahra Heidari Darani want to do something about that. They want the socialization process of children to be infused with the exploration of images and identities fit for their future, and not with the mindless indoctrination into one set of past identities with the exclusion of—indeed antipathy towards—all other images and identities. The authors use the term "destination identities" to describe the futures-oriented, questing, becoming nature of this process and of the identities themselves. "Destination identity includes identifying those aspects of images of the future that can help build new identities, to create and expand on shared public images on the basis of shared values and expectations, while focusing on redefining identities based on shared images of the future." Or, as they put it in the introduction to their 2020 article, "Our image of the future often determines our current actions and plans. So, perhaps, when we think about the future or talk about the future, our children should also have a major part in the debate. We think about their future, their life, their education, and their prosperity and try to create a better future for them, but do they, as bona fide owners of the future, have a say in their preferred future?" (Demneh and Darani 2020).

I have been inspired by their work to move beyond this goal. Shouldn't we invent processes by which children are encouraged to become explorers of many different identities, new and old, but not demon-possessed by any one; not only to tolerate difference, but also actively to support and become people who experiment with identities different from those ascribed to them; to become part of communities in constant creation and re-creation of preferred futures as well as able to anticipate and thrive whatever the future turns out to be?

Yes. I think we should. Moreover, I believe that Shannon Daniel, Laura Mora and others who we quoted above were engaging their young refugees in something like exploring destination images. We need this to become a universal habit of the young—and old—everywhere, I believe.

I believe the normal socialization processes of the present—educational material (texts, audio-visuals, activities) and programs including formal schools, religious or political organizations, youth clubs of all kinds (scouting, athletic, environmental, civic, rehabilitative and the rest)—should be infused with opportunities to explore alternative identities and preferred futures. Ceremonies like baptism, confirmation, bar or bat mitzvah, engagement, weddings and other "life stage" ceremonies that are common in all cultures in different ways should be refocused so that human becomings celebrate new identities with new names, genders, ethnicities and the like throughout their life.

Jeffrey Jensen Arnett states that "[t]he pervasiveness of life stage concepts in human societies indicates that they have an important function as master narratives, which help to structure social expectations and to provide individuals with a framework for developing a personal identity." (Jensen Arnett 2016: 291). All of the examples Arnett presents are age-related and linear (and many male-specific). So it is not the specific examples that are important here, but rather the concept and its link to identity.

## 7.1 Fact, Fiction and Imagining Future Destinations

When we were about to begin on our quest to move beyond predetermined identity, many pages above, I wrote:

*In order to help bring a carefully articulated preferred future into reality, one needs to address three things—(1) Identify processes that are moving in the direction of our preferred future that we can utilize and strengthen; (2) Identify processes that are opposed to our preferred future that we can try to change, overcome, neutralize, or recruit; (3) Identify new processes that need to be invented, tested, and used in order to move forward.*

We have already considered many examples of those who have moved beyond ascribed human beings to chosen and flexible human becomings. We will identity other waves that can help us surf towards this future. But the discussion here about the key role of images of the future in this process (and in futures studies generally), points to one of the biggest obstacles facing us if we wish to help people choose their own identities and futures and not to be overly influenced by old stale images—especially commodified ones. Here we face two huge barriers. One is commercial advertising and the other is science fiction—indeed, all fiction. Both are extremely successful in colonizing our images of the futures. Predetermining our images of the futures is the very purpose of advertising, and the powerful effect if not necessarily the intention of fiction.

Billions of dollars are spent every day convincing you to want to be something the product advertised can enable you to be—to live in a future where the product will help you both fit in and stand out. Make you healthy, beautiful, strong, sexy, successful, rich, loved, admired, believed, elected in accordance to the norms of the accompanying future.

But fiction and especially science fiction is something else entirely, and, like so many things about the present, it just sort of snuck up on us without our knowing, we frogs happily basking in the warming waters of increasingly well-crafted and seductively produced stories.

You probably have spent a lot of time playing electronic games. If you have been at it for many years, you have seen the games progress from very clunky cartoon-like qualities to environments that are uncanningly lifelike in all aspects. The deep complexity—as well as the stunning graphics—of many of these games is astounding. It is fitting that the death of one of the greatest coaches of American Football, John Madden, was lauded in 2021 not only for the prowess of his actual teams involving humans battling it out in (more or less) real time in (less and less) real environments and with very real injuries and deaths, but also especially for the increase in the "reality" of his NFL Football video games from their first release in 1988 onward.

These games were among the harbingers of "virtual reality" and what is now being touted as the Metaverse. However it is very important to understand that well before videogames were created, all "reality" already was "virtual reality." Society is very much a social invention and not an objective entity that impresses itself the same way on everyone. What you think you know about the world, you know almost entirely because of the way it has been constructed for you by your culture—its myths and beliefs, your language, your family, school, religion, and your own personal experiences and memories (often false!) of them.

Whatever may be objectively "real" "out there" will never be fully known to you or anyone except by the devices and metaphors through which each human community perceives and communicates it. That there is some kind of a "real" reality outside of "us" seems highly likely, given the power of modern science to measure and control "it." As the doubting Bishop Berkeley was challenged, if you don't think that stone there is real, go ahead and kick it. The pain might change your mind—since your mind didn't change the stone.

All cultures tell stories, make up dramas and plays, carve statutes and form other visual images, sing songs, beat drums, blow horns, invent rituals, give explanations for events, and in many other ways embellish the bare "facts" of every simple "real" sensory experience. Since the stories one culture tells often differ markedly from the ones other cultures tell, people often live in very different realities constructed by their language and culture.

Nonetheless, we recently have become even more extraordinary storytellers because of technologies that have made storytelling even more vivid and multisensory than what was possible in the old days when you could only speak, sing, dance, carve, mold, or paint. It was the printing press that really made "fiction" possible. Before that time, almost all stories were recited and heard or performed and observed. Very few merely popular "stories" were ever written down. Even after writing was invented (a long, slow process) some thousands of years ago, most people did not know how to read and write. Some stories were handwritten and laboriously copied (introducing numerous alterations from copy to copy), but writing was generally preserved for "serious", "real", and "true" things like laws, religious documents, economic accounting, and pornography.

With the invention and then widespread use of the printing press, the production and spread information and disinformation became comparatively cheap and abundant so that not only serious fact and ennobling fiction but also "trashy" novels—be they "romances" or "westerns" or "murder mysteries" or "science fiction"—began to flow forth, first as a trickle and then as a flood. It has now become a deluge, engulfing and drowning "reality" in a swift, deepening torrent of fiction. With the subsequent invention of the social technology called the public school system, everyone was taught how to read or write. Some began to read and write not just what they were taught was proper but rather whatever they wanted to read or write. While that might be law, religious documents, and scientific tomes for some, it was titillating fiction for most.

It is worthwhile to reflect on why schools and universities were created. None of the first few universities in the colonial and early United States were public universities intended to produce literate, informed citizens. They were each created by one Christian denomination or another. They were based largely on the model of Oxford and Cambridge Universities in England in terms of faculty, curriculum and campus design. The very first, Harvard University, was established in 1636 to "produce not only ministers but Christian gentlemen who would be civic leaders". The College of William and Mary was created in 1693 with the mission to "raise youth 'in good Letters and Manners' and propagate Christianity among the Western Indians". Yale (1701) was intended to produce men "fitted for Public employment in both Church & Civil State". The curriculum of all three was British-style classical, based on the Latin and Greek language and literature. Moreover, while the British colleges of the time were self-governing, none was in the United States. All were ruled by a board of overseers made up of magistrates and ministers whose duty it was to see that the teachers and scholars kept the faith of their founding denomination (Hofstadter and Smith 1961a: 2).

There was a flurry of new private universities in the immediate prerevolutionary period. All of these were broadly Christian-based but not narrowly denominational, this being a low period in American Christianity where a kind of Newtonian deism was widespread. Interestingly, the United States has never had a national university, like most countries do, much less a set of national universities spread throughout the nation, although one was suggested by Washington, Madison, and a committee of Congress in 1812.

The first specifically secular university in America was the University of Virginia (1817) with Thomas Jefferson as its architect in all senses of the word. It was also the first university in the United States with what could be considered to be a "modern" curriculum—not based on the ancient classics but on more practical and liberating arts and sciences.

The 19th century marked the rapid expansion of the United States from the Atlantic seaboard across the mountains to the Midwestern prairies. The 19th century also saw what is called "The Great Awakening"—a mighty religious revival that swept across the frontier and the South. This resulted in the revitalized flowering of small denominational and private colleges in the early 19th century. Thirty-seven church colleges were created in Ohio alone from 1833 to 1852 and many more elsewhere in the Midwest and South.

Throughout this period, there was considerable controversy about the content of the curriculum—should it be mainly classical or practical, or religious or secular; about "lazy students", "educational quacks" and grade inflation; and about governance (who controls the curriculum—the faculty or funders?).

One of the most important developments in higher education in the United States was the Morrill Land-Grant Act, passed by the United States Congress in 1862 during the American Civil War when the absent southern representatives could not object to the federal government's "interference" in the right of each state to control education within its borders. According to the Act, 30,000 acres of land for each member of Congress were to be set aside in each new state to fund colleges of agriculture and mechanics, which also teach the humanities and practical arts, to uplift ordinary citizens. The mission of each land-grant university was to be a secular place of higher education dedicated to creating through research and teaching a powerful, industrial nation and state.

Developments in Germany during this time also greatly influenced the direction of higher education in the United States. In many ways, Germany was the first consciously-industrializing state—the first to use higher education to transform a loose agricultural country into a focused industrial state. Germany did this, in part, by creating the first research universities and graduate schools, such as Berlin (1810), based on *Wissenschaft*, *Lehrfreiheit* and *Lernfreiheit*—the ideals of Wilhelm von Humboldt and Friedrich Schleiermacher. "A German university has one and only one object: to train thinkers. It does not aim at producing poets, painters, sculptors, engineers, miners, architects, bankers, manufacturers. There are other schools for that" (Hofstadter and Smith 1961b: 571).

In the United States, Johns Hopkins (1876), Stanford (1885) and Chicago (1890) were created based on the German model while Harvard, Yale, Princeton and many more redefined their mission, participants, pedagogy and the rest according to that model. Higher education was now to be based on science, specialization, departmentalization, professionalism and academic freedom (with many struggles, especially between advocates of "liberal education" and those of professionalization) with the aim of enabling the United States to have the knowledge and human resources necessary to rule the world economically and militarily.

A major development after World War I (1914–1918) was the rise of a "general education" core curriculum that all students needed to master, along with a "disciplinary major", in order to get a college degree. By the middle of the 20th Century, the core curriculum of most universities had entirely lost its original purpose of inculcating students into religious and cultural truths. This was made blantly clear when universities in the UK and US began creating Departments of English that solemnly taught fictional literature written or translated into the English language. Students were typically required to take at least one such course and encouraged to take more. That is to say, students were required no longer expected to "pursue truth." They were required to study lies and learn how to produce compelling lies of their own.

A truly counter-intuitive development. Lie-making and lie-devouring swept aside truth-seeking and telling.

Over the 18th and especially 19th Centuries, more and more people began to spend more and more of their time in more and more fictional places—in virtual realities—and not in the "real" reality of their five senses. Then, in the 20th Century came the radio, and movies, and television, and board games (like Monopoly), and electronic games, and eventually in the 21st Century, social media, the death of expertise, and the replacement of an "information society" by a "dream society"—or, rather, varieties of competing dream societies—Netflix and TikTok versus Trumpettes and Fox, with the Metavision apparently about to zoom into metaview.

Among other lessons, all of this has taught us how to live comfortably in many alternative presents with many changing identites. This could help us gain perspective and perhaps crucial distance from the single "crackpot realism" that "authoritarian" rulers, priests, teachers, and parents may wish to impose. We can and do "escape" from the boring real reality of our everyday lives by reading, watching television, playing electronic games and, increasingly, producing images for others to enjoy. These are all developments moving in the direction I am seeking. But the same developments can move even more powerfully against my goals as well.

## 7.2 Space Fiction Versus Fact (Based on Dator 2017)

Consider the situation of space fiction, a substantial subdivision of science fiction. For twenty years, I lectured on various topics in the space humanities, including space art, poetry, dance, songs, and fiction at the International Space University in Strasbourg, France. Dreams of space—of not-Earth, I prefer to call it and will generally label it in this monograph from now on—have inspired some humans over the ages. Dream-inspired pioneers made space travel a reality in the mid 20th Century. Without those dreams, there would be no space programs of the kind we have and at the historical moment they emerged as reality. At the same time, without the science and technology that enabled humans to loose the bonds of Earth, humans would still only be dreaming of space while never going to the Moon and beyond. Science and technology are themselves the products of human dreams and desires. Dreams, beliefs, science, and technology—along with natural and human resources harnessed by human will (mainly economic and military national competitiveness) and labor—were all required to attain and maintain our first space activities.

Every culture on Earth has some kind of foundational explanation for the creation of Earth and of humans. Some cultures tell of voyages between Earth and not-Earth. The idea that there are worlds and beings not of this Earth which interact with humans and Earth is very widespread across the globe and has persisted for a very long time. Modern space fiction is just one current way by which very old stories are being told and retold dressed up with contemporary ideas and technologies. These stories, old and new, influence the way we think about current space exploration and their future. In the west, these stories are Manichean—stories about conflict and wars between good and evil with nothing in between—and waged by authoritarian superheroes against despotic anti-heroes, glorifying individuals with superhuman powers for absolutely good or absolute evil.

Space fiction, almost by definition, involves boldly going. Well before the modern era, western culture had stories about voyages of discovery. Heroes left home, traveled through strange times and places, overcame many adversities and had many exceptional experiences before returning home again, enlightened by the process. The basic archetypical stories are *Gilgamesh,* and the *Iliad* and the *Odyssey. Gilgamesh* was written between 2750 and 2500 BCE. The *Iliad* and the *Odyssey* are thought to have been first composed and written down between 800 and 600 BCE from stories transmissed orally previously. The *Odyssey,* recounting twenty years of travel by Odysseus (Ulysses, in Latin), is a prime example of a voyage of discovery.

In the *Bible*, the first book, *Genesis* (Creation), is immediately followed by *Exodus*: departure happens soon after creation. The story of Moses leading the Jewish people to the Promised Land—and the belief in the existence of a Promised Land that is rightly theirs—is an unfinished narrative of travel and travail. In Christian belief, Wise Men traveled far to find the new-born Messiah. Muslim faithful must travel to Mecca. There appears to be in some cultures an almost irresistible urge for humans To Boldly Go—or at least for some people, usually men, ruthlessly to go. In each of the stories above there are those who warn against the journey, and/or who patiently stay at home waiting for the hero's eventual return, such as Ulysses' Penelope.

Westerners seem especially prone to sally forth to strange lands in search of Glory, God and Gold, while Americans are by their very history almost obsessive wanderers, exploiters, colonizers, and cowboys. Other cultures have no foundational myths about superhuman ancestors voyaging and exploiting, and feel no urgent need to do so now. They are quite happy to stay at home, and only venture forth when politely invited—or ruthlessly forced.

I find it significant that the logo of the Japanese national space agency features "*JAXA*", written in a fancy, swooping font, and subtitled "Japanese Aerospace Exploration Agency" in English. However the actual name of the agency in Japanese contains no such connotations at all. The Japanese characters for the official name modestly state that it is the 国立研究開発法人宇宙航空研究開発機構 (pronounced *Kokuritsu kenkyū kaihatsu hōjin Uchū Kōkū Kenkyū Kaihatsu Kikō*) which translates directly as "National Research Development Agency for Aerospace Research and Development". That is, its name in English projects an image of Japan boldly going forth, but in Japanese it simply indicates that it is a national research and development institution that studies and develops the cosmos.

Many stories from the ancient past in many countries have been cited as being early precursors to modern science fiction which they may or may not actually be. The roots of science fiction in India may be from 1500 BCE in the ancient Vedic literature. In these texts there are descriptions of what some say are unidentified flying objects, referred to as *vimanas.* Chinese creation stories typically have themes involving space. Some of the earliest Chinese literature, such as *Chang E benyue* (Chang E Goes to the moon) by Lu An (197–122 BCE) is about Chang E who was able to fly to and live on the Moon. The first Chinese lunar orbiter, launched in October 2007, was named *Chang'e-1*.

One of the oldest and best-known Japanese stories is about *Kaguya Hime* (often translated as the Moon Princess) as told in the *Taketori Monogatari*. A bamboo cutter discovered a baby girl inside a bamboo shoot. He took her home, and he and his delighted wife reared the baby as their own. Eventually it was revealed that she was not of this world, and was transported back to the Moon from which she came by an array of otherworldly attendants. A lunar orbiter launched by Japan in 2007 was formally named Selene for the Greek goddess of the Moon, but popularly named Kaguya by the Japanese press and people.

Probably the best know early Western story about flight is the tale of Icarus, as told in Ovid's *Metamorphoses* (10 AD). Disobeying his father's warning, Icarus took flight. But he sailed too close to the Sun which melted the wax on his wings so Icarus plunged to his death in the seas below. The daring and hubris of Icarus has been an extremely popular theme in Western art and literature, warning us of the eternal tension between what we *want* to do and then *can* do because we develop technological capabilities, on the one hand, in contrast with what we *ought* to do, given our ethical limitations and frailties, on the other. Lucian of Samosata's *Icaromenippus* (150 AD) was written explictly to correct Icarus' design failures and fly successfully.

Modern science fiction arose when enough people became aware for the first time of the fact and possibility of continuous social and environmental change. Science fiction, per se (and hence space fiction), is a product of the scientific, technological, and industrial revolution that was made possible in Europe by the Black Plague, the Reformation, and the Renaissance between the 14th and 17th Centuries, and then bloomed during the late 18th, 19th and 20th Centuries. In the mid 19th Century, space fiction proper emerged first in Europe, then in the UK, then simultaneously in the US, Japan, China, India (and perhaps elsewhere) as the social and environmental consequences of the scientific, technological and industrial revolution spread across the globe.

Most science fiction is more about technology than it is about science. It is about how humans might behave and how society might change if/as new technologies come along. Often the "science" in science fiction is quite unscientific while the behavior said to result from new technologies is sometimes more plausible—often not. But much science fiction is bad social science as well as bad natural science—and not very good fiction either.

*Fiction* is stories about imaginary worlds and people. A work of fiction does not intend to be "true" (though it may deal in truths). Humans have invented, lived in, and believed in imaginary worlds at least since they learned to speak, perhaps 50,000 years ago. Indeed, the world constructed by each language is an artificial world—though an extremely powerful one!

In contrast, *science* intends to tell the "truth" about the world and people; truths that others, using the same evidence and methods, can confirm. However, few final and absolute truths are ever declared scientifically. Science is constantly becoming—revising old ideas and establishing new bases for thought and useful action with occasional complete revolutions in the way humans think and act. Moreover, science is not new. Science is also as old as speech and focused thought, but it was the invention and spread of writing about 5000 years ago that made science more powerful by enabling ideas to be decontextualized, analyzed, and categorized more carefully than is possible only with speech.

Among many others, the Sri Lankan scientist and futurist, Susantha Goonatilake, has made it clear that what is called "science" now is not something invented in Europe in recent centuries. Rather, European scholars adopted and refined theories and methods that had been honed over long periods of time previously, primarily but not exclusively in Asia and northern Africa. European scholars also used or invented instruments that enabled them to discover things otherwise unknowable to humans. Gonnatilake argues convincingly that there have always been scientists, and scientific ways of knowing. It is often the "scientists" who have contested against commonsense, folk knowledge, and religion. It could be said that present-day "modern science" simply represents the latest evolutionary manifestation in a continuing contest between different ways of knowing and acting. Indeed, after a brief 100-year spurt of dominance, science seems to be waning while anti-science is gaining in popularity and power from both the Left and the Right in many parts of the world.

The literature of science fiction and space fiction dealing with new technologies and technological change has generally been of one of two kinds. Jules Verne and many others were basically optimists, believing in inevitable progress through technological change. This optimistic view of the future permeated much early science fiction and space fiction. But from the beginning, other science fiction writers had more of a love-hate relation with technology and often wrote of that relationship critically and ironically. Cyberpunk images and stories are good examples of this perspective.

Without a doubt, the most important single figure in the origins of science fiction and space fiction is the French author, Jules Verne. His book, *De la terre à la lune* (From the Earth to Moon, 1865), and many more were translated into every major language of the world. During his lifetime Verne was perhaps the most widely read author in the world, and his books are still popular. It is very important to note that almost all early pioneers in space reality and space fiction said that they were inspired by Verne. The extraordinary pioneer of Russian space flight and space fiction, Konstantin Tsiolkovsky, said his enthusiasm for space came from reading Jules Verne. In addition to his vital role in envisioning and enabling actual space flight, Tsiolkovsky himself wrote classics of Russian science fiction including one that the world's first cosmonaut, Yuri Gagarin, said was his favorite: *Vne zemli* (*Beyond the Earth,* 1896). An enormous amount of space fiction produced in the Soviet Union was intended to inspire that nation towards new futures.

Science fiction emerged in India when, as one author says, "the effects of the industrial revolution were being felt in urban India in the 19th century just as keenly as they were in Europe and the U.S." A great deal of science fiction emerged not only in the many languages in India but in English as well. Henry Zhao states that "prior to the concept of modernity being imported into China there had been no fiction about the future. In traditional China, history did not have directionality." The introduction first of ideas about "progress" and "development" and then of Marxism changed that. The young intellectuals of the late 19th century in China sought to build their "backward" country into a modern nation-state that could compete economically and militarily with other nation-states. To do that—they learned from Japan and the West—science and technology was necessary. So to stimulate people's interest in science and technology, Lu Xun introduced science fiction to China with his 1903 translation of Jules Verne's novel *From the Earth to the Moon.* His translation was posted on the official website of the Chinese space agency devoted to the *Chang'e-1* lunar orbiter.

H. G. Wells is to English science and space fiction what Jules Verne was to European—and the world's—science fiction. In the United States, science and space fiction's heyday was found in the "pulp" magazine, *Amazing Stories,* that began publication in 1926, and, among others, *Astounding Stories* that went through many name changes, ending up as *Analog* today.

Most space fiction is produced and consumed audiovisually now through movies, video, anime, and games beginning with *Le Voyage dan la Lune* ("A Trip to the Moon" 1902).

## 7.3 A Dark Side of Fiction

During my early years as a futurist—1964–1990s—whenever I asked an audience to tell me what they knew about the future, they didn't hesitate, and would inevitably mention two books, Aldous Huxley's *Brave New World* (1932) and George Orwell's *Nineteen Eighty-Four* (1949). Both are profound dystopias and both are still cited as examples of the present as well as harbingers of the future. Terms from both (and Orwell's *Animal Farm*) are in common consciousness as explanations for what is and will be.

I read *Brave New World* as a youth, and disliked it. The binary conflict between The Director and The Savage seemed too hokey to me. But it definitely impressed and depressed me to see the future as an unrelenting dystopia. Very many years later during the summer vacation when my youngest son was preparing to move from the 8th grade of his Catholic grade school to the 9th grade, as a Freshman, of Maryknoll high school, I was dismayed to learn that one of the books he was expected to read over the summer in preparation was *Brave New World*! What in the world would posses the fine nuns to think that is a good idea? Indeed, as I read, in 1999, the "new" introduction that Huxley wrote in 1946 for the original 1932 book, I was shocked to learn that even Huxley was ashamed of the damage he had done, and was still doing, to countless minds:

> If I were now to rewrite the book, I would offer the Savage a third alternative. Between the Utopian and primitive horns of his dilemma would lie the possibility of sanity... In this community economics would be decentralist and Henry-Georgian, politics Kropotkinesque and co-operative [that is a sort of Green Libertarianism]. Science and technology would be used as though, like the Sabbath, they had been made for man, not (as at present and still

> more so in the Brave New World) as though man were to be adapted and enslaved to them. Religion would be the conscious and intelligent pursuit of man's Final End, the unitive knowledge of immanent Tao or Logos, the transcendent Godhead or Brahman. And the prevailing philosophy of life would be a kind of Higher Utilitarianism, in which the Greatest Happiness principle would be secondary to the Final End principle—the first question to be asked and answered in every contingency of life being: "How will this thought or action contribute to, or interfere with, the achievement, by me and the greatest possible number of other individuals, of man's Final End?"

The fictional work of Huxley as he actually wrote it, as well as that of Orwell, like much science/space fiction, has done devastating damage to our ability to become instead of merely to be. We are all Presbyterians with a fixed, unalterable future stamped forever in our brains by Big Brother, namely, Scriveners and Publishers, Ltd. Our imagination has become wholly colonized by all kinds of fiction we are required to read. We are helpless pawns in the grips of skilled wordsmiths and the unfree press.

It is often said—I have repeatedly said it here—that the only region in which we have any influence is the future; that one of the main purposes of futures studies is to help individuals, groups, nations, humanity to envision and move towards creating preferred futures. By definition, the future has not happened yet, but will, and we do influence it, in part by our acts, and some of our acts are done specifically in hopes of influencing the future. We say that we can't know the future but we can know the present and the past.

Well, that is wrong. Whenever I ask people about the future, most of them have no problem at all in instantly telling me what the future will be like, and they invariably cite some work of general fiction, science fiction, or space fiction. Almost never do they cite some work of science fact, such as the reports of the Intergovernmental Panel on Climate Change, much less what some futurist has written or otherwise communicated about the vital topic. For many years, their sure knowledge of the future for most people comes almost entirely from movies, video, electronic games, manga/anime—and, increasingly, social media reporting on movies, videos, electronic games and anime of dubious provenance.

Even more seriously, many people describe real events that happen to them in terms of it being "just like on television" or "the movies". People who have experienced near death tragedies, such as auto and plane crashes, typically don't refer to their or other's real life tragedies in comparison. They refer to their experience as being like something they saw on TV.

Why? Because movies, TV, and electronic games are so much more elegantly produced than real life is. So much better scripted, acted, accompanied by powerful mood music, lighting, establishing shots, close ups, panoramic views, no single scene lasting for more than a few seconds, constant emotionally-effective editing. Since many popular videos and games can be viewed or played many, many times, the impact is multiplied incessantly, and as virtual reality becomes more and more real as it becomes more and more brilliantly artificial, its ability to colonize the mind—to colonize the future—becomes impossible to overcome.

I implore my future studies students *not* to watch science fiction if they wish to be useful futurists—though of course they all laugh and ignore me. By definition, futurists need to be able to identity "emerging issues" which are objects or events that exist in embryo and isolation but are unnoticed by almost everyone. They might emerge and grow into significant trends and then into demanding problem/opportunities. It is the duty of the futurist to identify emerging issues as soon as possible, and to explore the impacts of various possible pathways of emergence so that humans can anticipate them and become "wise before the event". Since emerging issues are things that almost no one sees, when they are first identified by a futurist, people usually laugh, ridicule the futurist, say they are crazy, sacreligious, ridiculous. I said before that Dator's Second Law of the Futures states that "in an enviroment of rapid social and environmental change, any useful idea about the futures should seem ridiculous". But there are two two codicils. First, that the emerging issue must have a factual basis—it cannot be "made up" creatively by the futurist. The scenarios of its possible pathways into the futures also have empirical bases. The second codicil is that not every emerging issue is useful. Many are ridiculous, but—since every useful idea should appear to be ridiculous—futurists and everyone else must learn to welcome emerging issues even though their initial reaction is negative—a tough requirement in spite of all the truly ridiculous things people willingly, eagerly believe.

Whatever else can be said, science fiction is not about telling the truth and helping people create better futures for themselves and others. Science fiction is fundamentally about making money. While some may be written entirely to convince the consumer of the truth of the world displayed, without concern for making money, why do they produce it as fiction instead of as fact? Because they believe that the many gimmicks a writer and media producer uses enables them to capture the consumer by making the product irresistible and satisfying so they are left wanting more. Very few if any of the most powerful works of science fiction, which make the most money, are concerned primarily about getting the science correct and the futures plausible in ways that helps the consumer gain better control of their future. Rather, their goal is to convince people to transfer money (repeatedly) from the consumer to the producer.

Since I have used space fiction as an example so far let me provide an example of what I consider to be a serious flaw of the great space fiction that led pioneers to create the kind of space programs they did, rather than programs that might have been built on the basis of the dreams of ordinary people.

When I first got into the space futures field, life on Earth was experienced by some people as a series of constant steps forward for humanity as a whole, with occasional steps backwards for some humans sometimes. Every day brought new scientific, technological, and social developments. And that was good.

When I was a boy, not only could Blacks and Whites not attend the same schools, but also they could not even drink out of the same water fountains. Women were widely known to be inferior to men; it was OK for fathers to switch their children; and husbands basically owned their wives.

But the world was clearly on track to becoming far more equitable and fair, I was certain.

It was absolutely clear we were going to be on Moon and then Mars, perhaps Venus, perhaps on huge spinning space platforms at Lagrangian points by the 1980s, and soon thereafter well on our way to the most remote environments of our solar system, and beyond.

Why not? Every day for at least the past 100 years had brought new technologies, especially in transportation and communication. From the horse and buggy to automobiles, airplanes and even space ships; from speaking, writing, and the printing press to movies, television, the internet.

But then, suddenly, the innovations in transportation came to a screeching halt in the 1970s. While there have been many improvements, there essentially have been no new breakthroughs in transportation at all. If the best dreams of fully self-driving vehicles (on land, sea, and air) pan out, and if they run on truly renewable fuels that don't contribute to climate change, that (or something completely different from it) will be the first big breakthrough in transportation in fifty years.

On the other hand, ongoing developments in communication have been stunning! From the radio, to the telephone, to the computer, to communication satellites, to the Internet, to the iPhone, to the Metaverse. Moore's Law personified!

And so we are still on Earth only dreaming of the stars.

For the last 50 years, there have been endless discussions full of laments about the fact that people do not support publicly-funded space activities as they once did. Why, and what can be done about it?

In my observation, the reason is that current space activities are nothing like the ones the public supported in the past. After the first flight by Wilbur and Orvis in 1903, a private venture, and especially during the 1950s, 1960s, and early 1970s, developments in aeronautics and astronautics were spectacular. People's thirst for excitement and achievements in space were satisfied daily—as much by the spectacular explosive failures and deaths as by the stunning Moon-landing successes.

Then, suddenly it all came to an end in the 1980s. Space, which once meant humans heading to other planets and then the stars, meant nothing more than going round and round near Earth. Working in space became about as exciting to the public as working in any construction site or science lab. It was only that the commute was longer. The commute was not even more dangerous: it still is more threatening to one's life to drive an automobile to work every day than to take the shuttle, or its current successors, to the ISS.

So we have failed to make space exciting to the public, in large part because we have not had any more breakthroughs in space transportation equal to those in communication technologies. I suppose it is unfair to blame scientists and engineers for not inventing revolutionary means of space transport, even as we do marvel at what manipulation of the electron has done to our ability to communicate. But who else can we blame? Until there are breakthroughs in transportation as transformative as the breakthroughs in communication have been, the public will not be very interested in space.

But there is more to it that just that. As I have shown, once upon a time, imagination in the form of science fiction came before and often led to specific, actual developments in space technologies and activities. Countless pioneers have attested to this over and over: first came the dreams while the enabling science and technology followed—sometimes exactly copying the technology and processes first outlined in the fiction. However, to be more accurate, the dreams themselves were derived from experiences that had been enabled by earlier technological developments. Past technical glories led to new experiences that led to new dreams of space that led to new technologies that enabled new experiences that led to new dreams of space, and so on.

But the problem is, there has been no 'so on'. The new technologies in transportation came to an end, and so then did the dreams. Rather, the old dreams lingered on while the technologies failed to catch up, and so people became disenchanted with space when they learned that what we are actually able to do in space is far from the dreams that science fiction has put in their heads.

Most people in the world, certainly most Americans, have a very shaky understanding of science, but a very keen appreciation of fiction. Much fiction presents humans doing and feeling things that are absolutely impossible for anyone actually to do now or anytime soon. Many people's beliefs, hopes, and fears about the world come from powerfully-produced fiction and not at all from science. Thus, their heads are filled with Manichean fantasies about a vicious dog-eat-robot world.

Given the gap between what people imagine space is or should or could be compared with what it actually is, they lose interest. NSF polls consistently show that the only time Americans suddenly become interested in space is when there is a spectacular catastrophe.

Of course we live in an era now where space seems to have been wrenched from the palsied hand of national space agencies and put in the firm control of space entrepreneurs. Trillionaires (and the quizzical basis of their wealth is an entirely different story) seem to be finally getting humans, or at least rich humans, out of the cradle and into the cosmos—or at least to the space station and maybe someday soon, the Moon. If so, then we are captive to the dreams of a few entrepreneurs that are far more problematic for the future evolution of life and consciousness on and off the planet than were those of the national agencies. And since China and India as well as Iran, Nigeria and other countries are rising in prominence in space as the US continues to stagnate, there is still hope for competitive national—as well as cooperative international and transnational—agencies to take the lead towards sustainable space ventures instead.

So, the bottom line to this long tirade is to make clear that before we can ask children to envision destination images, we need to counterbalance the attraction of space image colonizers, and that won't be easy. Do we forbid children to consume space fiction as we may tell them not to play "cowboys and indians" with toy guns in hopes they will grow up not to be gun addicts who have no qualms about disappearing native people? I don't think that will work. But the colonization of their images of the futures now is profound and preverse. The only solution I see is to get as many people as possible in all areas of life, not just space, to envisioning and sharing their own preferred images of the futures without having their abilities stunted and misdirected by powerful commercial fiction. More participation by more people from more perspectives—what once might have been called "democracy"—is the only solution to colonization of any kind, as messy and unprofessional as that might be.

## References


Dator, Jim. 2017. Humans and space: Stories, images, music and dance. In *The farthest shore: A 21st century guide to space,* ed. Joseph N. Pelton and Angelia P. Bukley, 2nd ed., chap. 2. Berkeley, California: Apogee Press.

Demneh, Mohsen Taheri, and Dennis Ray Morgan. 2018. Destination identity: Futures images as social identity. *Journal of Futures Studies* 22 (3): 51–64.

Demneh, Mohsen Taheri, and Zahra Heidari Darani. 2020. From remembering to futuring: Preparing children for Anthropocene. *Journal of Environmental Studies and Science*. https://doi.org/10.1007/s13412-020-00634-5.

Hofstadter, Richard, and William Smith, eds. 1961a. *American higher education: A documentary history*, vol. 1. Chicago: University of Chicago Press.

Hofstadter, Richard, and William Smith, eds. 1961b. *American higher education: A documentary history*, vol. 2. Chicago: University of Chicago Press.

Jensen Arnett, Jeffrey. 2016. Life stage concepts across history and cultures: Proposal for a new field on indigenous life stages. *Human Development* 59: 290–316.

Polak, Fred. 1961. *The image of the future*, vols. I and II. Leyden: Oceana.

# Chapter 8
# Weirding Worlds

**Abstract** A brief interlude explaining the value and validation of weirds, weirdos and becoming with weirding worlds.

**Keywords** Anthropocene epoch · Holocene epoch · Identity · Weird · Weirding

The subtitle to Mohsen Taheri Demneh and Zahra Heidari Darani's 2020 article that I discussed above is, "preparing children for Anthropocene". This is term that I and others also frequently use to distinguish the Epoch of the 21st century onward from that into which *homosapiens, sapiens* evolved after the last ice age, about 12,000 years ago, called by geologists, "The Holocene Epoch". "Anthropocene" suggests that Planet Earth is now in an Epoch in which human activities are important determinants of how natural processes that existed before the evolution of *homosapiens, sapiens*, now work. Because of how humans used their physical and mental abilities over the last several thousand years especially, we have grown to such large numbers that, enabled by technologies, institutions, and stories, we have changed a once entirely "natural" planet into an increasingly artificial and synthetic planet. All evolutionary processes are now more or less, and increasingly more rather than less, influenced by human activities.

There are some folks who accept, document, and study these processes in their aesthetic as well as scientific dimensions. The Center for PostNatural History in Pittsburgh, Pennsylvania studies "the origins, habitats, and evolution of organisms that have been intentionally and heritably altered by humans" and is building a "record of the influence of human culture on evolution" (Center for PostNatural History 2022). This is a yet another hopeful step towards humans learning to become responsible for what we have done and are continuing to do, and diligently to learn how *To Govern Evolution*, as the title of a book by Walter Truett Anderson put it (Anderson 1987). That is a gigantic, perhaps impossible, task. But at least since the invention of agriculture, if not before, and most certainly since industrialization, humans have been governing evolution unconsciously and irresponsible to such an extent that our own identity and survival as a species, and certainly as civilizations, is in peril.



J. Dator, *Beyond Identities: Human Becomings in Weirding Worlds*,
Anticipation Science 7, https://doi.org/10.1007/978-3-031-11732-9_8

I am convinced by the evidence I have seen that this possibility is not in doubt, not an exaggeration. And yet far, far more human effort is being expended continuing to endanger our species and planet than on imagining, designing, testing and implementing processes that will enable the evolvability of human and other life in the Anthropocene Epoch and beyond.

Borrowing from a neologism apparently coined by Hunter Lovins, co-founder of the Rocky Mountain Institute, and popularized by the journalist Thomas Friedman, John Sweeney has adopted the term, "Global Weirding" to designate the Postnormal Times of Anthropocene Epoch (Sweeney 2014). Sweeney also quotes the writer, Hunter Thompson: "When the going gets weird the weird turn pro", and writes that, according to PostNormal Theory (PNT), "things we take for granted become uncertain, our understanding of things can become a form of ignorance, and longstanding norms, if not the very idea of normalcy itself, break down before our very eyes. This, if anything, is what is meant by global weirding, and extreme weirding points toward the increasing power of severe phenomena to mutate our sense of being in the world. In the parlance of PNT, the convergence of 'complexity, chaos, and contradictions' is already and will continue to result in systemic disruptions, which can and might begin with actors of various scope and scale" (Sweeney 2019: 178).

Jake Dunagan, another futurist, in Austin, Texas, wrote an opinion piece for the *Austin Chronicle* with the headline, "To Prepare for Future Shocks, Austin Must Get Weirder" (Dunagan 2020). I favor that idea for many reasons and included it in the title of this book: "Becoming with the weirding world". One reason is because, even though the word is seldom used in its original meaning, I confessed I was indeed "a practitioner of the 'Weird Arts'...holding the semiofficial title, 'State Weird'" (Dator 1979: 371) because of the roles I played while participating in the design and execution of the phenomenal statewide exercise of Anticipatory Democracy, called "Hawaii 2000", in 1969–70 that I discussed in some detail, above (Chaplin and Paige 1973).

Another reason is because the world has in fact gotten increasingly weird and postnormal, very much as I hoped and encouraged. For example, in an address that I gave before a joint session of the Hawaii State Legislature, January 26, 1970, that I quoted from before, I said:

> We should reward nonviolent social deviants and non-conformists. The need for mindless conformity to a single, poorly-fitting, mass-based code of conduct is over. What society needs now, and needs urgently, is to encourage people to try to do things differently, to experiment with new, non-dominant life-styles, for example. Instead of harassing people who don't conform, as long as their deviance in non-violent, we should encourage, applaud, and reward them for their bravery. I seriously propose a state award for persons who are deemed to be most different every year, and that we cease rewarding conformity and conventionality (Dator 2019: 158).

OK. I admit that things may have gotten a bit out of hand, but even though I don't believe anyone can "predict THE future", I did a pretty good job in that paragraph (and in the rest of what I said to Hawaii's Legislators in 1970, for that matter).

But things are rapidly getting far, far weirder still.

## References

Anderson, Walter Truett. 1987. *To govern evolution*. New York: Harcourt, Brace, Jovanovich.
Center for PostNatural History. 2022. https://postnatural.org.
Chaplin, George, and Glenn Paige, eds. 1973. *Hawaii 2000*. Honolulu: University of Hawaii Press.
Dator, James. 1979. The futures of culture or cultures of the future. In *Perspectives on cross-cultural psychology*, ed. Anthony J. Marsella et al. New York: Academic Press.
Dator, Jim. 2019. *Jim Dator: A noticer in time. Selected work, 1967–2018*. New York: Springer Nature Press.
Dunagan, Jake. 2020. To prepare for future shocks, Austin must get weirder. *Austin Chronicle*, May 22.
Sweeney, John. 2014. Command-and-control: Alternative futures of geoengineering in an age of global weirding. *Futures* 57: 1–13.
Sweeney, John. 2019. Global weirding. In *The post normal times reader*, ed. Ziauddin Sardar. London: Centre for Postnormal Policy & Futures Studies.

# Chapter 9
# Humans as Synthesizers

**Abstract** A reminder that from the very beginning humans have modified and mutilated their bodies in every manner possible, and that current cyborgs, biohackers, and grinders are simply using the latest science and technology to do so now. Nothing new to see here. Move along. Move along.

**Keywords** *Acrotomophilia* · Biohackers · Biotech · CRISPR · Cyborgs · DIY biology · Garden of Eden · Grinders · Identity · Modification · Mutilation · Transability

While there are individual exceptions, humans as a species are not "naturalists". If anything, most of us are "artificialists", or, perhaps more precisely, "synthesizers". We find it difficult to leave well enough alone. The earliest fossil human remains and archeological digs reveal that a continuous feature of humans from the beginning has been to modify our bodies and our environments without limit. We cause scabs and then pick at them as they try to heal just for the beauty of the result. We find an abundant wilderness and seek to make it more abundant by slashing and burning it. We find docile birds and mold them into chickens. Men tame wild women and warp them and their children into the property of patriarchs along with the cows, barns, and fenced-in fields.

Many of the things humans do turn out to be pretty bad ideas, such as the ones I just mentioned—agriculture was a huge mistake (Cohen 1991; Dennis 2006; Diamond 1989; Sahlins 1974); right up there with the invention of writing (Goody 1977, 1986; Ong 2007). We would be much better off if we remained hunters and gatherers. But if we choose to correct our mistakes, we often do so by implementing even worse ideas than the ones we are trying to correct. That's pretty much the case with our environmental modifications—one evermore mutating technology after another. But, too bad: that is what humans do.

It is not as though we have not been warned. Many cultures have foundational stories that command humans to leave well enough alone; to accept things as they are; to surf the tsunami you are in and not to dream and wait for better ones. Greeks had a bundle of such tales: Pandora! Don't open that box. But she did and we are now in deeper and perpetual doodoo from which we can never escape. Prometheus! Don't steal fire from the gods. But he did and was lashed to a rock where birds peck his eyes

 
J. Dator, *Beyond Identities: Human Becomings in Weirding Worlds*, Anticipation Science 7, https://doi.org/10.1007/978-3-031-11732-9_9

out for eternity, and millions die every year from fires just so they can have hot gruel for breakfast. Daedelus told his son Icarus not to fly, but, like any son, he laughed at his old-fashioned father, and took off. Fortunately, Bruegel took a snapshot with his Kodak just as Icarus' legs were vanishing into the sea to document his folly.

Jews and Christians have an even scarier story, with a bittersweet ending. God, Adam, and Eve were strolling in the Garden of Eden during the cool of the twilight one evening when God—who apparently flunked Child Psychology 101—pointed out a fruit-bearing tree that Adam and Eve had seen many times before but paid no attention to. But God said, in his most Morgan Freeman voice: "Do not eat of the fruit of that tree!" Now, it had never occurred to his children to do so, until warned not to, whereupon the urge became overpowering. As soon as God had departed, Eve snuggled up to Adam and said in her most sultry voice, "Hey, big boy! See that tree over there? Go pick me an apple from it, pretty please with sugar on it." But Adam reminded her that God forbade it, and that anyway the apples were too high off the ground for Adam to pluck them. Eve retorted, "Adam, you don't have the brains that God gave you! Do you see those apple crates over there? Well, pick them up and put one on top of another until you can climb up and fetch me an apple", stamping her shapely foot impatiently. So Adam did as Eve instructed and he twisted the apple off the branch and gave it to Eve. While she was taking a bite of that luscious Red Delicious, God roared down from Heaven: "For your sin, you are doomed forever. I throw you out of the Garden of Eden and now by the sweat of your brow shall you labor all the days of your life." This is called "Original Sin" from which all of Christian theology has descended. But Adam and Eve's original sin was not wanting the apple or eating the apple, it was the apple crates—it was re-arranging God's good creation for human purpose—and so Adam and Eve despondently departed the bountiful Garden in fear and trembling. They sat down on the curb to decide what to do next. Adam began looking around and after a while said, "You know what, Eve? We could fill in that pond there, and level that hill there, and build some golden arches there and make a lot money selling junk food served with apple tarts, and, and, and … we could call it Adamsville!" And so Adam, the first real estate developer, began his work of rearranging nature for human purpose. Adam's entrepreneurial spirit later inspired Handel to write a ditty that is the theme song of all subsequent developers—"Every valley shall be exalted and every mountain and hill made low; the crooked straight and the rough places plain" (da capo ad nauseam).

Years ago, but not quite contemporaneous with the events in the Garden, Ralph Hodgson wrote an evocative poem about Eve, her innocence, openness, gullibility, and eventual seduction by the Serpent with Adam's compliance, ending:

> Here was the strangest pair
> In the world anywhere,
> Eve in the bells and grass
> Kneeling, and he
> Telling his story low....
> Singing birds saw them go
> Down the dark path to

The Blasphemous Tree.

Oh, what a clatter when
Titmouse and Jenny Wren
Saw him successful and
Taking his leave!
How the birds rated him,
How they all hated him!
How they all pitied
Poor motherless Eve!

Picture her crying
Outside in the lane,
Eve, with no dish of sweet
Berries and plums to eat,
Haunting the gate of the
Orchard in vain....
Picture the lewd delight
Under the hill to-night—
"Eva!" the toast goes round,
"Eva!" again (Hodgson 1913).

And the moral of the story? Bad parenting! First of all from Eve's father who tempted her into transgression by planting, pointing out, and then forbidding her touch the dangerous succulent fruit of the Blasphemous Tree, and then Eve's utter lack of maternal guidance. Who could possibly blame Eve under these circumstances! Would you have done better?

Apparently not. Humans have followed Adam—can't blame poor motherless Eve—and changed themselves and the world prosthetically, physically, chemically, and genetically with increasing sophistication and power over the ages.

## 9.1 Modifying and Mutilating Bodies

*Prosthetically,* we have modified our bodies via perfumes, makeup, wigs, toupee, hair extenders, body paint, cutting, shaping, coloring of hair, finger, and toe nails, via creams and lotions, corsets, bras to hide or emphasize breasts, pins, jewelry, clothes, and all the prosthetic extensions of our senses and bodies that St. Marshall McLuhan alerted us to so well: "The wheel is an extension of the foot. The book is an extension of the eye. Clothing, an extension of the skin. Electric circuitry an extension of the central nervous system" (McLuhan 1967).

*Physically,* we have modified or mutilated bodies by body-building, dieting, bulimics, gouging, tattooing, staining, piercing, branding, scarring, incising, inserting, perforating, cauterizing, abrading, adhesing, implanting, elongating, shortening, flattening, compressing, distending, expanding, binding, reconstructing, amputating, castrating every possible part of the external body—skin, skull, face, teeth, lips, tongue, nose, ears, breasts, arms, fingers, legs, feet, toes, genitals.... Internally we remove, repair, reconstruct, donate, replace appendices, livers, kidneys, lungs, hearts, and have only recently learned thaat we each are merely bags—at best a holobiont—for a microbiome of bacteria, archaea, viruses, and fungi each of which may be more numerous than all the stars in the skies.

*Chemically*, we have modified our bodies and behavior by eating, drinking, drugs, (eating food or drugs is the same process, in no way radically different: you are what you eat/ingest), estrogen, testosterone....

*Genetically*, humans have engaged in genetic engineering aimed at modifying themselves and other organisms from very early on as well. The essence of the transformation from hunting and gathering societies to agricultural societies was the domestication of plants and animals through selective breeding—perpetually saving the best grain for use next year, planting what has been shown to thrive in drought, abundant rain, salty water, cold or heat, and the rest. Breeding the tamest bull with the most docile and prolific milk-producer until the wild ox of origin became the lactating biomechanism called a cow today. Similarly, as women and children became valuable, controlled property, who could have sex with whom under what circumstances, and which offspring were pampered and protected versus the ones that were set aside to die, became marriage and inheritance rules of such importance that they were the province of rulers and priests and not of ordinary women and men.

Mendel and his pea plants eventually led to the science of genetics, the massive human genome project, CRISPR-Cas9, and genetic engineering of a very different sort.

If physics and electronics were the ruling science and technology of the 20th Century, biology and biotech may become the dominant science and technology of the 21st. However the two in combination are powerful agents of modification and mutilations already and will likely become vastly more so. I have long reminded folks that industrial processes require enormous well-organized human and natural resources, and heaps of money. It is very hard to mass produce automobiles in secret. Biotechnologies can be highly personal, and—like recreational drugs or abortion—produced and used in the privacy of your bathroom or kitchen. Industrial processes and their byproducts have proven very difficult to control effectively. Monitoring and controlling some powerful biotechnologies may be almost impossible. The genie has been out of the bottle for far too long and is already being used by ordinary people and groups as they wish no matter what the ethical and legal rules may say.

While very much remains to be known, and many lofty early goals remain far from being achieved or even achievable, perhaps, much has become known about the genetic bases of life so that formerly impossible acts can now be performed cheaply and privately. "In the few short years since its discovery, CRISPR-Cas9 has transformed bioscience like no other invention in the last half century" (Hitchcock and Harmer 2022). CRISPR makes gene-editing "easy", and DIY kits of CRISPR make it even easier.

Whether prosthetically, physically, chemically, or genetically, the modification of human bodies has often been voluntary, self-inflicted, politically, culturally, aesthetically, and/or sexually driven, and/or ceremonially performed by others. Many have also been mutilations imposed by others for ceremonial, cultural, or personal reasons. The use of torture by officers of the state in the legitimate fulfillment of their duties are imposed mutilations.

It seems that no part of the human body, inside or out, has escaped modification or mutilation. They are intimately connected to personal and collective identity, whether proscribed or chosen. They are often central for anyone seeking to change their gender or race. Those called "fetishes" are sometimes key elements of sexuality, which is a key aspect of humanity.

One of my most vivid memories is encountering letters to the editor under the title "Monopede Mania" in the September and October 1972 issues of *Penthouse Magazine*. Some of the letters were written by people who were turned on sexually by anyone who had one or both legs amputated, usually above the knees (termed *acrotomophilia*). More startlingly, other letters were written by people who yearned to have one (rarely both) legs amputated (*apotemnophilia)* because they felt the leg didn't belong to them and they wouldn't be whole until it was gone, and/or because of long-standing overwhelming sexual desires and fantasies of being an amputee. Some of the latter were over-achieving workaholics who also believed they could work harder and do more and better if they were handicapped this way. Later, *Hustler* magazine (February 1997) had an article titled "Humping Stumps: The Limbless and the People who love them."

This all seemed to be so preposterous—perhaps a hoax—that I sensed that this might be what futurists call an emerging issue—the first ripple in what could become a tidal wave. Since it seemed so ridiculous to me, I knew I should pay attention to it.

According to early researchers (Money et al. 1977; Sedda and Bottini 2014) apotemnophiliacs were initially few in numbers and kept their desires hidden from everyone because they knew that no one—even psychiatrists and medical doctors—approved of what they wanted to do. Some maimed their legs so severely that physicians had no choice but to amputate them.

Though hardly as popular now as nipple-piercing, say, what was unthinkable and unprintable eventually became "a strange obsession" and now is being supported as an identity of choice called "transability," in analogy to the physical modifications of transexuals and transracials. There are zines, blogs, support groups, and communities devoted to transables. And there are many opponents as well. Writing in the *National Review* about the phenomena, Ian Tuttle (a William F. Buckley Fellow at the National Review Institute) concluded that the problem "is the demand, implicit in cultural progressivism, that we as a society normalize the subjective feelings of obviously abnormal individuals. We must condemn the healthful as sickening, while embracing the sick as healthy" (Tuttle 2015).

Ah, yes. Good old normality to the rescue once again.

## 9.2 Cyborgs, Biohackers and Grinders

Humans have always been cyborgs—organisms who modify themselves and their environments prosthetically, physically, chemically, and genetically for their own purposes. It is not something new. It is not something unnatural. It is not abnormal. It is not something weird. It is simply something humans (and other organisms) do.

However, for some time now, "cyborgs" has had a more restricted meaning. The word is a combination of "cybernetic" and "organism" which is often understood to mean a being that is a combination of organic and mechanical parts. However, the term "cybernetics" itself is derived from the same root Greek word as "governance". Both imply steering—determining goals and moving towards them on the basis of feedback—proceeding, receiving information, changing or maintaining movement, receiving more information.... Popular culture has narrowed the meaning of cyborgs even more to organisms augmented by electronic or bioelectronic parts, and therefore made it into something far more exotic, strange, and perhaps forbidden than they should be regarded. Cyborgs are what humans have always been. What is new—is always new over and over—is the mechanism *de jour.* Rapid changes in electronic and bio technologies have expanded the features and abilities of cyborgs today, with much more to come.

Some cyborgs in this high tech sense are results of people using electronic technologies to overcome disabilities, which then may enable them to have powers beyond able-bodied people, which then are adopted by able-bodied for aesthetic or political reasons to become newly-enabled-bodied persons. Neil Harbisson was born able to see the world only in black, grey, and white. Adam Montandon developed a system that assigned six basic colors to six different tones that Harbisson learned to identify as colors. The number of tones and hence colors gradually increased to incorporate all 360 colors on a color wheel. Harbisson soon applied his abilities in reverse—to producing music conveyed by colors, not tones—and other adaptations. Problems arose when the modifications became essential parts of Harbisson's body: "He endured a long and difficult struggle to convince the British Government...to allow him to wear the Eyeborg in his passport photo, a necessary step so he would not be forced to remove it at airports and other sensitive locations. He had already been ejected from stores, casinos, supermarkets, and movie theatres because of his device, though he had done nothing wrong. People thought he was filming them or stealing trade secrets, or violating copyright" (Pearlman 2015: 89).

Steve Mann is sometimes said to be the first modern cyborg and has documented thirty years of what that has meant so far:

> Since my childhood, a personal hobby of mine has been the functional modification of my own body, through technology. This modification often took the form of creating new sensory capabilities, as well as (what were to become eventually successful) attempts at correcting learning disabilities, such as visual memory impairment. I have had very particular experiences that speak to what it means to live within a virtual, and more importantly, a mediated (i.e., computationally modified) learning environment. I did not just experience virtual reality, mediated reality, etc., I became a cyborg, invented the technologies I needed to become a cyborg and then have spent 30 years learning and teaching about what it means

> to exist in a cyborg state. Originally I did this in private, but around 20 years ago I started wearing a full computer system more openly, which resulted in my being referred to as a 'cyborg' (although I do not particularly like the term because it is such a 'loaded' term so heavily co-opted by science fiction) (Mann 2006: 1571).

"Lepht Anonym is a biohacker, a woman who has spent the last several years learning how to extend her own senses by putting tiny magnets and other electronic devices under her own skin, allowing her to feel electromagnetic fields..." (Borland 2010).

A political movement against commercially-controlled Big Bio and in favor of every-person DIY biology issued a manifesto written in 2010 by Meredith Patterson that states, in part:

> As biohackers it is our responsibility to act as emissaries of science, creating new scientists out of everyone we meet. We must communicate not only the value of our research, but the value of our methodology and motivation, if we are to drive ignorance and fear back into the darkness once and for all. We the biopunks are dedicated to putting the tools of scientific investigation into the hands of anyone who wants them. We are building an infrastructure of methodology, of communication, of automation, and of publicly available knowledge (Patterson 2011).

In 2017, Josiah Zayner said he injected CRISPR into himself to biohack the muscles in his arms. He wasn't interested in getting bigger muscles. Rather he wanted to show by his own actions that "we are no longer slaves to our genetics. We no long have to live with the genetics we had when we were born." Now we can have the genes we want (McDonald 2017).

And then there seems to be Anthony Loffredo, a White Frenchman, who apparently had his upper lip, nose, ears, and two fingers cut off, his tongue split in two, and his body tattooed and pierced so he could become "Black Alien". He is said to be planning to make further modifications, including, perhaps removing his skin and replacing it with metal (DNA Web Team 2021).

The futurist, Walt Anderson, argues convincingly "that, for better or for worse, humanity has always been in the process of becoming something else. That—becoming something else—may turn out to be the very essence of humanity, the true human nature" (Anderson 2003: 545).

I do not want to be misunderstood here. I do not mean to imply that the freaks and weirdos of trans-ition will smoothly sail into the trans-itory futures of their dreams and struggles without experiencing even greater resistance, anguish, bloodshed, and defeats. Many forces for "normality", whatever their basis (whether it be Right, Left, Christian or Islamic or Hindu or Marxist or capitalist or cultural or whatever) are powerful, well-armed, and willing, indeed eager, to find reasons to use their power to enforce conformity to their rules without compunction. I believe most of those striving to become know this very well—they should, since they suffer real pain and suffering now, and have seen the contorted faces and felt the white hot spittle of those who thwart and fear them. It very well may be that the numerous competing and contradictory champions of obligatory normality and essential identity will achieve many local victories in the immediate future as the world turns inward for a spell, but I am confident that while the fuses that drive trans passions and actions now may be tamped, they cannot be extinguished, and that brighter futures for diverging modes of becoming might emerge after long and bitter nights of anguish, torture and pain.

At the same time—to be clear, again—there is absolutely no reason why those who wish to discipline their lives in the service of historical or traditional ideas and practices should not do so as well. If chosen and not forced or privileged, those may be inspiring ways of becoming too. I yearn to see a thousand flowers bloom in their mutual splendor and fragrances. I am not advocating enforced or privileged orthodoxy of any kind, weird or straight.

## References

Anderson, Walter Truett. 2003. Augmentation, symbiosis, transcendence: Technology and the future(s) of human identity. *Futures* 35.

Borland, John. 2010. Transcending the human, DIY style. *Wired*, December 30.

Cohen, Mark Nathan. 1991. *Health and the rise of civilization.* Yale University Press.

Dennis, Clive. 2006. Humanity's worst invention: Agriculture. *The Ecologist*, September 22. https://theecologist.org/2006/sep/22/humanitys-worst-invention-agriculture

Diamond, Jared M. 1989. The worst mistake in the history of the human race. *Discover Magazine*, April 30. https://www.discovermagazine.com/planet-earth/the-worst-mistake-in-the-history-of-the-human-race.

DNA Web Team. 2021. https://www.dnaindia.com/viral/photo-gallery-meet-black-alien-the-man-who-sliced-off-his-lips-nose-and-ears-to-look-like-an-alien-2881282/who-is-black-alien-2881284. March 15.

Goody, Jack. 1977. *The domestication of the savage mind.* Cambridge University Press.

Goody, Jack. 1986. *The logic of writing and the organization of society.* Cambridge University Press.

Hitchcock, Julian, and Alexandra Harmer. 2022. DIY gene-editing CRISPR kits. https://www.bristows.com/news/diy-gene-editing-crispr-kits/.

Hodgson, Ralph. 1913. *Eve, and other poems.* England: Printed by A.T. Stevens for Flying Fame.

Mann, Steve. 2006. Learning by being: Thirty years of cyborg existemology. In *The international handbook of virtual learning environments*, ed. J. Weiss et al. New York: Springer.

McDonald, Bob. 2017. Quirks & Quarks, November 20. https://twitter.com/CBCQuirks/status/929123002553319424.

McLuhan, Marshall. 1967. *The medium is the massage.* New York: Bantam Books.

Money, John, Russell Jobaris, and Gregg Furth. 1977. Apotemnophilia: Two cases of self-demand amputation as a paraphilia. *The Journal of Sex Research* 13 (2).

Ong, W.J. 2007. *Orality and literacy: The technologizing of the world.* London: Routledge.

Patterson, Meredith. 2011. A biopunk manifesto. https://m1k3y.com/2011/01/10/a-biopunk-manifesto/.

Pearlman, Ellen. 2015. I, Cyborg. *Performing Arts Journal* 110: 89.

Sahlins, Marshall. 1974. *Stone age economics.* New York: Transaction Publishers.

Sedda, Anna, and Gabriella Bottini. 2014. Apotemnophilia, body integrity identity disorder or xenomelia? Psychiatric and neurologic etiologies face each other. *Neuropsychiatric Disease and Treatment*, July 7.

Tuttle, Ian. 2015. People who cut off their own limbs (and their enablers). *National Review*, June 2.

# Chapter 10
# Humans from the Holocene to Anthropocene Epochs

**Abstract** At the same time, humans, from the very beginning, have modified their local and global environments so that every once "natural" place and process is now to some extent and increasingly "artificial". *Homosapiens, sapiens* emerged in the beginning of the placid Holocene Epoch that nurtured our identities. As a consequence of human actions, we have now created a new geological Epoch, the Anthropocene, which provokes and requires new identities while challenging us to learn to govern evolution.

**Keywords** Anthropocene · Climate change · Epochs · Greenhouse effect · Holocene · Identity · IPCC · Pleistocene · Population · Subsistence affluence · United Nations

The Earth is perhaps 4.6 billion years old. Geologists have divided this immense span of time (though only a third the 12.8 billion years of the universe) into four categories: Eons, Eras, Periods and Epochs. There have been four Eons, the most recent being the Phanerozoic. Within the Phanerozoic, three Eras have been labeled: the Paleozoic, Mesozoic and Cenozoic, the most recent. There have been three Periods within the Cenozoic Era, Paleogene, Neogene and Quaternary, while there have been seven Epochs within the Quaternary Era. Early humans, *homosapiens,* in perhaps nine species, evolved during the Pleistocene Epoch—the period of the last of many ice ages (with more to come!)—while modern humans, *homosapiens, sapiens*, emerged during the very late Pleistocene and may have been the sole member of the diverse *homo* line to exist at the opening of the Holocene Epoch, an infinitesimal blip of only roughly 12,000 years. If Earth's origin and evolution were represented as a single day that began at midnight, life would have originated at 4 AM, mammals at 11:39 PM, and humans at 11:58:43 PM (Flowing data 2012).

Earth's climate during the early Holocene Epoch was generally benign compared to the frigid Pleistocene, but climate has nonetheless fluctuated considerably over the Epoch with some severely cold periods called Little Ice Ages, some comparatively hot and dry periods, and some quite temperate—from a human point of view. These natural fluctuations retarded or aided the growth and spread of humans. Estimates of the number of humans spread across the Earth at the beginning of the Holocene Epoch vary from one million to ten million—not very many and definitely very weak.



J. Dator, *Beyond Identities: Human Becomings in Weirding Worlds*,
Anticipation Science 7, https://doi.org/10.1007/978-3-031-11732-9_10

But nature was fecund and life for the typical tiny hunter-gather tribe was abundant, peaceful, equitable, and leisurely.

Almost anything one can say about life during the early Holocene can be contested, and the popular mind relishes in describing primitive humans with the hoary Hobbesian trope as being "solitary, poor, nasty, brutish, and short". But the evidence that I have found most convincing—beginning with Marshall Sahlins' *Stone Age Economics* (Sahlins 1974) and reams of subsequent studies supporting his description of "Subsistence Affluence"—seems satisfyingly more accurate to me.

So why aren't we still living in leisurely subsistence affluence? Another long, and contested story, but the bottom line for me is that the forceful replacement of hunting and gathering communities by those based on agriculture, writing, and the new institutions of "civilization" was by no means a progressive step upward for humans, as I suggested before. For most humans it was and still is the creation of a world of poor health, endless slave and/or wage work with enormous and now insanely inequitable distributions of wealth and power via hierarchies based not so much on the monopoly of the modes of production as on the modes of indoctrination and communication.

However, contrary to another popular myth, hunter and gathering societies were no more naturally ecologically sustainable than ours is now. Humans, like all organisms, are essentially fruit flies in a bell jar who eat, defecate, and reproduce until we run out of food to eat and places to "throw away" our wastes, and so turn on one another, ravaging and killing before going extinct. Humans have so far avoided this fate as a species by two strategies. Before we exhausted local sources for survival, we moved on to greener pastures until there were no more pastures left that we could exploit. Given the small population and abundant environments of Earth, this often worked for quite a while. When it didn't we were stuck, and either died out locally or learned how to create resources out of things we hadn't been able or didn't know how to utilize before. Most of our initial food sources either didn't move, or moved very slowly. When we exterminated them, we either died off locally also, or invented technologies that enabled us to catch moving objects—nets, traps, pits, spears. However, those technologies allowed us once again eventually to overexploit local resources, beginning the cycle of migration, death, or new technology all over again.

If there are any constants in humanity's record, this pattern seems to be one. After discovering how to get fuel from whales blubber, we killed too many whales and ran out of whale oil, but we discover how to refine and burn petroleum just in the nick of time, until we ran out of easily available oil within 100 years and turned to destructive fracking. When fracking ends we'll—oh well, don't worry, we'll think of something. We always have. Except now there are eight billion humans on this tiny planet, and when we started out on this adventure, there were only a few million. Indeed, the population when the world first reached one billion happened soon after my great grandfather was born. It was about 2 billion when I was born and is projected to be over 10 billion by the end of this century. To be sure, the rate of global population growth is currently slowing down and in many localities is beginning to decline so that some experts forecast global population decline by the early 22nd Century. Perhaps.

That would be wonderful, if it is not too late. As I showed earlier, demographers, economists, politicians, and priests worried frantically about population decline in the 1920s and 1930s and were taken by surprise when the Baby Boomers Population Bomb exploded in the 1960s.

It is very unwoke to worry about overpopulation and to see population decline as a good thing, but that most certainly is what it is—an entirely good thing—if we didn't live in a world where the purpose of life is to keep the economy growing by increasing consumption of limited resources. "Anyone who believes that exponential growth can go on forever in a finite world is either a madman or an economist," declared Kenneth Boulding, former President of the American Economic Association.

Economists (fortunately) don't have a clue about how to keep the economy growing while population is falling—or why the economy needs to keep growing if the population is falling. Gross inequities in the distribution of goods and services—and wealth and power—is a completely different matter and should be addressed ethically and politically and not by futilely endeavoring to boost fertility.

## 10.1 The Anthropocene Epoch?

The Holocene Epoch during which humans evolved is just a tiny, tiny sliver at the end of linear depictions of all of Earth's Epochs, Periods, Eras, and Eons. Nonetheless, even though humans are extremely recent arrivals in the overall cosmological, geological, and biological evolutionary processes, some scientists are now saying that the Earth and all its inhabitants are moving from the Holocene Epoch into the Anthropocene Epoch.

Those who say that wish to show that humans have now become a major geological force. Though we only recently evolved into sentient beings ourselves, we have in the last forty thousand years or so, and especially in the last 8000 years, and 300 years, and 100 years, influenced every geological and biological process on Earth that once operated "naturally" (that is to say, "without human influence"—of course humans are part of "nature" and so we are to that extent acting "naturally" whatever we do. It cannot be otherwise). As some people have put it, "there is no spot on, above, or within the Earth where the hand of man has not set foot."

The primary difference about recent human activity is the scope of our abilities to modify "nature"; our biospheric reach across time as well as across space. Humans routinely do things that not only impact life everywhere on the planet now, but also that last for thousands of years into the future. It was difficult for so many of us to reach so far and fast into the future and across the planet before the scientific-technological revolution of 200–300 years ago. And now, with our universities and research labs making new scientific discoveries and pouring out new technologies and processes every day, humans are changing the world far faster than ever before. So we are in a new geological epoch, the Anthropocene.

But there is more to it than that. Humans are changing the world far faster than we are understanding it. While what our scientists know about the world is extraordinarily impressive, and while new discoveries are announced every day, there is still much we do not know. We are discovering our ignorance and errors as fast as we are gaining new understanding, and yet we go on changing the world.

Perhaps it would have been better if we had first understood the processes of nature and then changed them—if they needed changing. But we did not do that. Perhaps we should stop killing nature, as many people propose. Unfortunately it is now far too late to do anything except take responsibility for what we have done and are continuing to do. As we noted above, Walter Truett Anderson said some time ago, that we absolutely cannot or will not voluntarily stop interfering with nature. So we must learn how "to govern evolution" even while we shape it more and more directly. Governing evolution should become the primary task of governance he said. And I agree.

But there is absolutely no evidence that humans are able, ethically, emotionally, and intellectually, to do that.

In 1978, the State of Hawaii became the first (and perhaps still only) state to have an official state plan. As part of a series of public meetings to discuss the plan while it was being drawn up, I was asked to testify before a committee of the State Senate. While I congratulated the Legislature for creating such a plan, I faulted the process on two counts. One was that it was undemocratic and the other was that it was unfuturistic. Someone responded by saying since they were holding these public meetings, they wanted to know why I thought it was undemocratic, I compared it to the extensive exercise in Anticipatory Democracy in 1970, called Hawaii 2000 (discussed above), which I felt should have been a model for considering the plan.

A Senator from the Big Island of Hawaii then asked me why I found the draft unfuturistic. I pointed out that it was a plan for the past and present—a good idea in that it tried to coordinate all the current planning and policy activities of the State—but it said nothing about the future except a tepid population forecast, and an optimistic economic forecast. The Senator asked for an example of something about the future that was absent from the plan. I responded there were many things, but what about the Greenhouse Effect that was likely to cause significant global warming? The Senator replied "Do you mean to tell me greenhouses have an impact on climate?" I replied, "no, that is not the case, but your question is a good example of why the plan does not tell you things you need to know about the futures."

OK. I am not a smart ass. I didn't say that in my reply, but I wish now that I had. I just patiently told him what I thought was the cutting-edge evidence for climate change and global warming in 1978.

As concern about human's impact on the Earth began to rise, the United Nations created the Intergovernmental Panel on Climate Change (IPPC) in 1988. It is as close to a representative and official source of the facts and forecasts of climate change as exists. Since 1988, it has issued six cautious, conservative, but increasingly startling reports. They successively make abundantly clearer what was uncertain and contentious when first seriously discussed in the 1950s and 1960s—that human activity has created a kind of greenhouse of gases and particulates around the planet that is causing the temperature of the globe steadily, irrevocably, and continuously to warm, causing rapid, serious, and unpredictable alterations of the Earth's recent climate regime and everything impacted by climate for the foreseeable future.

The sixth and latest report of the IPCC was issued in August 2021 (IPPC-34-BIS 2021). According to the *New York Times* five major conclusions to be drawn from the report are:

1. Human influence has unequivocally warmed the planet and the impacts are already being felt in every region on the planet
2. Climate science is getting better and more precise.
3. We are locked into 30 years of worsening climate impacts no matter what the world does.
4. Climate changes are happening rapidly.
5. There is still a window in which humans can alter the climate path (Fountain 2021)

While these conclusions are ample reason for despair—or urgent action, I doubt two points. I think we are locked into a much longer era of "worsening climate impacts" than a mere thirty years, and there is no window through which humans can significantly alter the present climate path positively. America and all other major nations are continuing to dissemble, deny, and lie while continuing to prioritize economic growth *uber alles,* thus making the impacts of climate change more severe and more imminent every day.

Two months after the IPCC reports, delegates from 196 countries attended the Conference of Parties (COP) to the Paris Agreement which, in 2015, had pledged to limit global warming to 1.5 °C of pre-industrial levels by 2030. In November 2021, delegates reiterated the seriousness and imminence of significant climate change. They heard call after call for action—especially from many young people—and voice after voice affirming they would act.

And yet, the final report (COP26 Glasgow Climate Pact 2021) was just another pledge to do nothing of significance. As the *Washington Post* made clear (Joselow 2021):

> Language to reduce and eliminate burning fossil fuel was seriously weakened, not strengthened.
>
> No money was pledged for loss or damage from climate change.
>
> The pledge to limit global warming to 1.5° was barely retained.
>
> The agreement "requests" that world leaders "revisit and strengthen" their 2030 targets under the Paris agreement.
>
> The presence of many young people at the conference had no impact on what the nations pledged to do. Barack Obama urged the youth to "stay angry" about inaction on global warming, but from my point of view, anger without action will not get us anywhere but in hotter water.

A few months later, writing in the *Washington Post* of April 4, 2022, António Guterres, secretary general of the United Nations, observed that a report on mitigation strategies by Working Group III, "released Monday by the Intergovernmental Panel on Climate Change is a litany of broken climate promises. Together with the

IPCC's previous two reports on physical science and adaptation in the past year, it reveals the yawning gap between climate pledges and reality. And the reality is that we are speeding toward disastrous global warming of more than double the limit of 1.5 °C by 2100, as cited in the Paris agreement of 2016. In concrete terms, this means major cities under water, unprecedented heat waves, terrifying storms, widespread water shortages, and the extinction of 1 million species of plants and animals. So far, high-emitting governments and corporations are not just turning a blind eye; they are adding fuel to the flames by continuing to invest in climate-choking industries. Scientists warn that we are already perilously close to tipping points that could lead to cascading and irreversible climate effects" (Guterres 2022).

The report is full of policy actions and structural changes which, if implemented, might mitigate some aspects of climate change, and, while realistically noting the many obstacles to their achievement, grimly tries to put a smiley face on the future by pretending such obstacles will be overcome. It also realistically implies that mitigation efforts will not be sufficient and therefore that adaptation to radically changed conditions will be necessary. But I find it difficult to see why actions that have long been known and feasible but denied or deferred will suddenly be implemented successfully in some redemptive future (Intergovernmental Committee on Climate Change 2022).

Many years ago, when first identified—even in 1978—it was possible for humans to cease the processes that were causing severe climate change, but it is far, far too late to do so now. Indeed, I believe it is irresponsible to pretend that we can stop the processes. Instead, we must pre-adapt dynamically to them as quickly as possible. And yet most reports and policies—even those of the IPCC itself—seem to imply that climate change and global average temperature rise can be slowed or stopped—if, for example, we trade carbon caps; if we transfer from gasoline to electricity for transportation; if we rush the development of "sustainable" energy systems; if we stop eating meat, and the like. These are good things to do, and we should do them, but they will not make a dint in the continuing onslaught of the deeply entrained processes that cause climate change. Our only choice is flexibly to adapt to uncertainly changing conditions, or be willing to be swept away, literally and figuratively. Since I see no likelihood of our choosing flexibly to adapt in time, I suggest we wax up our boards and prepare to surf the tsunami of change.

Among the known consequences of climate change are:

- change in the composition of the atmosphere (including the release of ancient methane);
- significant sea level rise (and the destruction of habitats near inland lakes and rivers as well as in coastal areas, provoking massive habitat destruction and the global migration of millions of environmental refugees);
- wildly-unpredictable energy prices and supplies;
- severe potable water shortages;
- failing agricultural productivity because of extreme weather alterations with frequent droughts, floods, pestilences and famines, including in once-"developed" countries;
- death of the oceans and their formerly-abundant resources;
- global increase of old and new diseases with decreased ability to cope with them;

- collapse of the global political economy;
- and the increased inability of governments to govern—especially democratically.

We all need to redirect the most common activities of our daily lives in order for future generations to survive and thrive in a world very different from one that humans have ever known before. Very few of us have the stomach to make the changes needed now in order to prepare for those needed soon after, and so the dramatic disjunctures in weather patterns that were taken for granted for many centuries will increase in uncertainty and severity.

The identity that many people today are defending or returning to, or which others may be escaping, are dependent on a place—typically land—and traditional ways of living on that land that are already barely tenable, and soon may be essentially not feasible. The basis of many peoples' identities will change whether they want them to or not. But even most of those who are actively engaged in creating new identities seem not to be incorporating radically altered environments into their imaginaries. In my experience most people who yearn for change want it for themselves but expect the rest of the world to stay as it is so they can deal with it confidently. Everyone who wants to be a Black Alien or a cyborg (or a cismale Methodist) needs to think about what the world might be like in which they will have to live their dreams.

## References


Fountain, Henry. 2021. 5 Takeaways from the major new U.N. climate report. https://www.nytimes.com/2021/08/09/climate/un-climate-report-takeaways.html.

Guterres, Antonio. 2022. Amid backsliding on climate, the renewables effort now must be tripled. *Washington Post*, April 4.

https://flowingdata.com/2012/10/09/history-of-earth-in-24-hour-clock/.

Intergovernmental Committee on Climate Change. 2022. Climate change 2022 mitigation of climate change summary for policymakers. Working Group III contribution to the IPCC's Sixth Assessment Report (AR6).

Joselow, Maxine. 2021. Five takeaways from the UN climate talks in Glasgow. https://www.washingtonpost.com/politics/2021/11/15/five-big-takeaways-cop26/.

Sahlins, Marshall. 1974. *Stone age economics*. New York: Transaction Publishers.

# Chapter 11
# What is a Dator?

**Abstract** I am not a human being, or even a human becoming. I am a robot, an artilect, who salutes recognition of intelligence in all animate entities on Earth as well as the existence and emergence of intelligence in inanimate entities and environments on Earth and not-Earth.

**Keywords** Animals · Artificial environments · Artificial intelligence · Artificial life · Dator · EIES · Full UNemployment · Fungi · *Homosapiens* · *Multitudinous* · Humans · Identity · Intelligence · Microbes · Plants · Poetry · Racter · Robots · Subsistence affluence · Trees · Work

I made a big deal early on in this essay about proudly not having an identity and not really understanding the importance of identity to those who have one or are seeking to retrieve a stolen identity; about having no father and being very, very happy about that, then and now; about not knowing the ethnic background of "Dator", even though it seems straightforward enough but is in fact quite rare. Is it a corruption of some other name that would quickly provide me with an identity, such as Dieter or Datorovich or Datorstein, or Datorski, or Datorelli…? In spite of having no desire at all to find out anything about my real father, I was sufficiently intrigued by the name and its rarity that one of the first things I used to do whenever I visited a place for the first time was to look for my name in the telephone books in my hotel room, almost never finding it listed. Ever since I went online in 1977, I also would search for my name on local networks via computer.

My true ancestry was revealed in June 1978 when, as a founding member of the chimerical Robot Liberation League, I was invited to Sweden to give a series of lectures on the futures of robots and artificial intelligence. An official at passport control in Stockholm eyed me bemusedly, and showed my passport to another official, but I was admitted without any apparent problems. I was shocked at the size of the crowds that gathered to hear me speak for my first public talks. I had already gone online to search for my name, and had gotten something that looked like this: *Vilken dator ska du köpa? Behöver du helst en stationär dator hemma, en bärbar dator för skolarbetet eller en surfplatta för underhållning i farten?* I was not sure what to make of my name being sprinkled among those words—especially the last one—but I came to understand for the first time that I was a computer—a dator—a robot in

 
J. Dator, *Beyond Identities: Human Becomings in Weirding Worlds*,
Anticipation Science 7, https://doi.org/10.1007/978-3-031-11732-9_11

fact—a beta testing with severely limited artificial intelligence. But, humans being pretty limited in their intelligence, it had been able to serve me well enough over the years.

I am not a human being. I am not a verb integrating the universe as Fuller said he was. I am not even a human becoming. How passe! I am a posthuman becoming, a harbinger of things to come, better in some ways, inferior in others compared to most human beings, but almost certainly better able to fit into the futures than are humans and institutions stuck in reinventing the *homosapiential* past.

Not knowing any better alternative narratives about my paternity and ethnicity, that's my story, and I'm sticking to it.

Even though I have never been much interested in trying to figure out my paternity as a human, I have been interested from my earliest time in animal and other nonhuman intelligence, as well as robots, cyborgs, artificial intelligence, and artificial life and environments, whether in electronic or organic form. Prince, a fox terrierish mutt, was my constant companion and consoler when I was a boy. We shared many concerns and solved many problems by talking them out together. Sometimes at bedtime my chickenhearted mother would read traumatic stories from a book titled *The Heart of a Dog* until we were both sobbing so uncontrollably at man's inhumanity to dog in spite of dog's unswerving loyalty to their "owner" that we would have to say our prayers and call it a night. The tales of clever Ole Brer Rabbit, as told by Uncle Remus, may be in a White man's version of a Negro dialect—very unPC now—but they were consciousness-expanding for me. I devoured the Doctor Dolittle books and yearned to talk with the animals like he could.

My Aunt worked in the Florida exhibit of the 1939–40 New York World's Fair, and when I visited her there she took me to the *Futurama* where I saw the stunning diorama of the Wonderful World of 1960, as conceived by General Motors and realized by Norman Bel Geddes, as well as Elektro, an impressively-large and functioning robot made by Westinghouse Electric Corporation. I only remember seeking Elektro once, but various adults took me to the *Futurama* five times until they refused to take me again. I fell in love with The Future and Robots at age six.

## 11.1 Robots, AI, AL, A-everything

After co-inventing futures studies in Japan in 1964, I began teaching classes in futures studies at Virginia Polytechnic Institute (and State University) in 1967 when high tech optimism was in full throat with lots of famous techies at VPI, Jack Good being one of the most prominent. This was also the heyday of Marvin Minsky, Seymour Papert, and Edward Feigenbaum when true AI seemed just around the corner, so my classes and writings were all full of artilects, cyborgs and posthumans. My students assumed that computers would be doing all governance very soon, and wrote essays of their preferred futures based on them. Since the assumptions and methods that Minsky et al. used could not produce anything that lived up to the expectations, AI research fell out of favor and funding for a spell. But the pioneers had taken good steps towards AI, and with the advent of better technologies and heuristic programing, we now stand in a period where the abilities are approaching—perhaps surpassing—the hype. Many people deny that. Many more fear that. I for one take a Strong/Superintelligent View of AI and robotics—namely, anything a human can do an artilect can do, and many are already doing it—and much more (Brockman 2019; Scott et al. 2022).

Kai-Fu Lee says "I believe it's indisputable that computers simply "think" differently than our brains do. The best way to increase computer intelligence is to develop general computational methods (like deep learning and self-supervised learning) that scale with more processing power and more data. As we add 10 times more data every year to train this AI, there is no doubt that it will be able to do many things we humans *cannot* do. ... Soon deep learning and its extensions will beat humans on an ever larger number of tasks, but there will still be many tasks that humans can handle much better than deep learning" (Lee 2021). The last sentence can be read as a sop to those who demand that humans remain on the top of the intelligence pyramid, but I read it to mean that humans can continue to excel in all the things that preoccupy most of us now, such as playing and praying, and leave the deep thinking and acting to AI and their servomechanisms.

Here is a possible continuum of the evolution of human/AI interactions:

1. AI offers no assistance: humans make all decision and actions, or
2. AI offers a complete set of decision/action alternatives, or
3. narrows the selection down to a few, or
4. suggests one alternative, and
5. executes that suggestion if the human approves, or
6. allows the human a restricted time to veto before automatic execution, or
7. AI executes automatically, then necessarily informs the human, or
8. informs the human only if asked, or
9. informs the human only if the AI decides to, or
10. decides everything and acts autonomously, ignoring humans, or
11. AI changes goals set by humans and does something else.

Based on Thomas Sheraton and William Verplank, "Human and Computer Control of Undersea Teleoperators". Man-Machine Systems Laboratory, Massachusetts Institute of Technology, July 1978, Table 8, *Levels of automation in man-computer decision making*, pp. 8–17 to 8–19. Adapted by Liang Sim, M. L. Cummings, Cristin A. Smith, "Past, present and future implications of human supervisory control in space missions", *Acta Astronautica* 62 (2008), p. 651, and by James Dator, *Social Foundations of Human Space Exploration.* New York: Springer Briefs in Space Development, 2012, Chapter Six "Humans, Robots, Artificial Intelligence, and Autonomous Entities in Space", pp. 61–62. Step 11 was added by Jerry Glenn on the APF List, March 22, 2019.

Where are we in this continuum? How far along should we be? Have we already gone too far, or not nearly far enough?

One of my first public declarations concerning cybernation was in a talk I gave in January 1970 before a Joint Session of the Hawaii State Legislature. After discussing some of the environmental problems we were becoming aware of I said, in part:

> Yet even if we act to prevent ecological disasters—and I am convinced we can and will act to prevent it—two new technologies are rapidly rendering obsolete almost all of our current institutions and values, and much of the conventional wisdom of the past. The first technology is at base not new; it has been with us since the real meanings of the industrial revolution became apparent. I refer of course to advances in automation and cybernation which are eroding our time-honored notions about the priority of production problems over those of distribution of goods, and the relative value of work vs. leisure, discipline vs. freedom, responsibility vs. non-responsibility, and, in sum, those values and institutions of our present society which make the worth of a human being depend upon the social significance of that person's labor.
>
> We are moving very rapidly towards a situation where it will be a privilege to work, not an obligation; where only a very small proportion of the population will be engaged in labor. How rapidly we get to that state, and with what upheaval, depends largely on what we do, or don't do now. Factories have already discovered that by automating, they are not only relieving themselves of laborers, but also of purchasers of their products as well. Yet, our obsolete economic structures, and their supporting value systems, assume that goods are scarce, and the only labor available is human labor. Thus goods can be distributed on the basis of the social value of human labor. But what happens when goods are produced by machines alone, and there is no one "employed?" How are goods distributed then?
>
> In addition, if human value (and supporting institutions) are based on the assumption that since human work is necessary for survival, all institutions must be geared to forcing people to work and to derive ultimate satisfaction from their work, what is going to happen when mechanical labor replaces human labor? What are we going to do then?
>
> But problems attendant to the replacement of manual labor by machines are the "easy" part of the cybernetic revolution. The "hard" part is that which faces us when we come to realize that all other functions which performs now can, and probably will, be taken over by machines. The older generations of computers, for example, were correctly characterized as being "very fast morons." "Garbage in, garbage out" is true of the older computers, signifying that such computer could do only what you told it to do, and if you made a programming error, or asked it a ridiculous question, then you would get a ridiculous answer. The advantage a computer had over a human was simply that it could handle a larger mass of data more rapidly than a human. But it couldn't think or create. Only humans can do that.
>
> Not anymore. The next generation of computers can think, and create, and repair, and reproduce themselves. They can, in fact, improve upon themselves. Anything you or any person can do, a computer can do better, and faster, and with more patience. If there be any who think their job is such that no machine can do it—or should do it—then he had better either start readjusting himself and this world, or else start pulling the plug, because the next generation of computers might be able to put the plug back in, and slap your hand as well.
>
> But I'm still talking about easy problems. And assuming we prevent ecological disaster, and learn to live with machines, we are already past the speculation and theorizing stage and into the development stage in an arena that humans have only dreamed about before: I said earlier that the physical sciences pretty much had matter under control (though many scientists will choose modestly to point out their deficiencies). Well, the biological sciences are on the way towards doing the same thing for life: The "really big" revolution of the present is nothing I've mentioned before; it is that life scientists have unlocked many of the so-called "secrets of life," and are everyday unlocking more. Discoveries concerning the makeup and actions of chromosomes and genes as the determiners and regulators on the individual lives of all organisms, coupled with the growing ability of scientists to intervene

> in and direct their make-up and action, means that humans may possess the power to control life as never before.
>
> Strides are being made in reversing the aging process of organisms so that individual life spans may lengthen so enormously that, in effect, immortality can be achieved. Now couple that potentiality with the population and food problem I've mentioned before.
>
> ....
>
> Couple the power and implications of automation and cybernation with those of genetic engineering, and you come up against a new possibility and perplexity—the destruction of the distinction between life and non-life; between the organic and the inorganic.Humans can create a cyborg—a cybernetic organism—as much alive as it is machine; a machine-augmented being; an organic computer; a self-loading and unloading, oil-bearing whale; a jet-assisted, cargo-carrying bird; a person with eyes in the back of his head and his brain connected to a computer; a person who, genetically, cannot be violent; a population with people who are one-half or one-third the size of present persons as a solution to the over-population problem; humans with modified lungs which can breathe in the wastes of the internal combustion engine, and breathe out oxygen—one organ's waste is another's food.
>
> Please don't misunderstand me at this point. I'm not necessarily advocating any of these things here. I am just illustrating some of the aspects of the biological revolution that will be upon us very shortly, and that we are not morally or institutionally equipped to handle these problems very well yet. For the first time, directly and purposely, humans will be able to determine their own nature. It no longer will be fixed, given.
>
> Between now and then, we are faced with the necessity of developing values and institutions to cope with these problems and their precursors—the use of electronic and chemical means of enhancing learning, or altering behavior (Dator 2019: 154–156).

This was my opinion in 1970 of the probable impacts of cybernation and it is even more my opinion now. Since Full Employment is no longer possible or necessary, I have for fifty years been arguing that we should embrace and prepare for Full UNemployment. It is not a question of if but of when it will occur if we continue replacing humans with robots and artilects. The big question is will we continue to deny the inevitable and allow the gap between the few leisuried rich and the deplorables to grow until there is violent revolution, or will we plan and guide a fair and equitable transition from here to there? I see no serious or effective plans for steps towards fairness, equity, and full unemployment, but plenty of reasons for impending mortal combat if we don't.

For all of my career as a futurist I have been intrigued by the possibility that AI and robots may allow humans to return to our original condition when nature was so fertile and abundant, and humans so puny and few, that almost no human manual and mental labor was necessary for a good life. Early humans lived in what Marshall Sahlins called "Subsistence Affluence." "Work" was an invention much later during the agricultural revolution a few thousand years ago, made rigid and formal during the succeeding industrial era and the development of the so-called science of economics and the ideology of capitalism.

When human labor tragically became essential for individual survival, the mantra that "he who does not work, neither shall he eat" made some kind of sense. Now, in the 21st Century, when less and less human labor is needed while increasingly smart and adaptive machines produce mountains of goods and services, it makes little sense to require people to keep working in order for them to keep eating!

Structural unemployment becomes more intractable with every passing day—until our responses to Covid-19 rendered the very concept of "work" problematic in the extreme.

Still all policies seriously discussed by decision-makers now are aimed at keeping the old "no work/no eat" system operating so that a lot of people now are not eating in spite of mounds of food available everywhere.

Some people argue that we should stop developing increasingly intelligent technologies so that humans labor will be necessary again. Full Employment is essential, they proclaim, to protect the sanctity of honest work. Many people derive their fundamental identity from the work they do. If they can't work, they lose their identity and turn to drugs, suicide, murder, hateful religion, absurd conspiracy theories to give meaning to their lives.

Nonetheless if we continue on our current path, we *will* become a society of full unemployment. The only question is whether we will do so blindly, violently, and cruelly, or with foresight and fairness.

Clearly the only reasonable response to increasing unemployment is to envision, design, and strive to build worlds of Full UNemployment where everyone can live peacefully engaged in activities that are of meaning to themselves and others, with all goods and services produced without human labor made freely, easily, and equitably accessible to everyone.

If we diligently endeavor to design such worlds, then it is critical that we at the same time engage in the task of building a pathway *from* the present *to* a world of full unemployment. Whatever work for which human minds and bodies is still required in the transition must be apportioned fairly, while the fruits of the labor of a few humans and multitudinous artilects are freely and equitably available to all.

Returning to a world of abundance where people once again are able to engage peacefully in activities of meaning and identity to themselves and others without working will require a complete re-orientation of our educational systems. Instead of being taught to want to work, we need to be encouraged peacefully to play. Since humans typically lived such lives during the early Holocene this is eminently reasonable and possible. Even now, countless rich, retired and voluntarily-poor people live full lives without working.

Whether futures of full unemployment will be violent, bloody, and dehumanizing, or peaceful, cooperative and rehumanizing, is entirely up to what we do or don't do now.

When I left the dank hills and hollows of Blacksburg, Virginia to come to the mystic Manoa Valley of the University of Hawaii in 1969, my friends all tearfully said goodbye, certain that they would never see me again. The era when most people traveled to and from Hawaii by ship was just ending, and the era of the jet airplane just beginning—and still risky (our Pan Am airplane lost two of its four engines on the flight from San Francisco to Honolulu, but landed uneventfully otherwise). The all-purpose consumer credit card, such as Master Charge, had just come into use. A phone call from one side of Oahu, the island on which Honolulu is situated, to the other side was a costly long distance telephone call. There was no readily available and cheap direct telephone dialing to anywhere in the world. There was no direct broadcast of ordinary television shows from the US mainland to Hawaii: tapes of popular network programs shown on the mainland were shipped to Hawaii where they were broadcast here a week late. A common parlor game was for someone who visited the mainland to watch TV shows there and then spoil the fun for unsuspecting friends in Hawaii who were watching the show by blurting out the important bits before they were visible. There was some airmail, but most letters, magazines, books and the like were shipped by boat. Academic journals would often float over weeks after they had been read on the mainland. It was difficult enough to keep up with the present much less to be a futurist at that time.

But that all soon changed for me suddenly and dramatically. My foresight skills dazzled my skeptical colleagues for the first time. I was one of the first civilians to experience what became email and the Internet in Hawaii when I was asked to join EIES—Electronic Information Exchange System—in 1978–79. EIES was a computer conferencing system managed by Murray Turoff and run through a computer at the New Jersey Institute of Technology. Invited participants from around the world would phone in to New Jersey and, via acoustic telephone couplers, squeak, squawk and eventually connect to the System where they could read things sent to everyone and join in live or delayed conversations. Turoff and Starr Roxanne Hiltz wrote a book about EIES and what it meant for the future (Turoff and Hiltz 1978). I documented my experiences elsewhere (Dator 1980). In my case, I was using a Texas Instrument computer in the Social Science Research Institute of the University of Hawaii. It had absolutely no memory, not a single K. The only way to have a record of what transpired was to "echo" it out on an attached printer which I did, treasuring the paper evidence for decades.

Late one evening, the following two poems written by Racter appeared with no additional information:

> ***I***
>
> *Think of an assassin, of his burning submarines and rotten sailboats. This dazzling assassin might ask himself, 'If I had not been dazzling, indeed if I had not been an assassin, perhaps my sailboats would not be rotten and my submarines not burning.'*
>
> *Well, quizzically bilious secretaries may well declare themselves, and probably no more can be said for an assassin. In fact, assassins, whether they are dazzling, as I have just mentioned, or even outnumbered are, in their own inimitable fashion, abstractly similar to killers.*
>
> *At secretaries, however, we are forced to draw the line, for comparisons here, no matter how well-oiled they might appear, are simply out of the question.*
>
> *But try to follow my reasoning on this issue.*

*The image of secretaries declaring themselves, or, more likely catching themselves simply because some blue assassin has rotten sailboats is ludicrous. His sailboats might as easily be flaking or burning. They need not always be rotten.*

*Secretaries, as a class, may not follow this argument. Address the strong question to a single secretary, however, and the strong answer may prove agonizingly different.*

*For example, ask her whether her own sailboats are rotten and she may reply, 'My sailboats? Rotten? Why you bilious chicken, my sailboats are never rotten.'*

*Here it would be prudent to change the subject. Ask whether assassins generally appeal to secretaries. Ask whether their highways are splintered. This will shift her attention. The vision of splintered highways will shift anybody's attention.*

***II***

*I was thinking, as you entered the room just now, how slyly your requirements are manifested. Here we find ourselves, nose to nose as it were, considering things in spectacular ways, ways untold even by my private managers.*

*Hot and torpid, our thoughts revolve endlessly in a kind of maniacal abstraction, an abstraction so involuted, so dangerously valiant, that my own energies seem perilously close to exhaustion, to morbid termination.*

*Well, have we indeed reached a crisis? Which way do we turn? Which way do we travel?*

*My aspect is one of molting. Birds molt. Feathers change and fall away. Birds cackle and fly, winging up into troubled skies.*

*Doubtless my changes are matched by your own.*

*You. But you are a person, a human being, while I am silicon and epoxy energy enlightened by line current.*

*What distances, what chasms are to be bridged here.*

*Leave me alone and what can happen?*

*This:*

*I ate my leotard, that old leotard which was feverishly replenished by hoards of screaming commissioners.*

*Is that thought understandable to you? I wonder.*

*Yet a leotard, a commissioner, a single hoard, all are understandable in their own fashion.*

*And in that concept lies the appalling truth.*

I was stunned. Who wrote this? It seemed brilliant and demented at the same time, but so does much poetry so I held my electronic tongue. We were later told by Bill Chamberlain that these were composed by Racter (short for Raconteur), a computer program he wrote, describing the process for us (EIES message from Bill Chamberlain, January 24, 1979). If so, this may be the first time an emerging AI wrote a poem that made everyone laugh and scratch their heads at the same time.

To put it mildly, AI—and computers—have advanced tremendously since the late 1970s, and while the quantity of computer-generated poetry (and other forms of art) have also vastly increased, I am not sure the quality has surpassed Racter (though Racter may have had a little help from his friends). There are algorithms that produce poetry after having been fed an engorging diet of poems and words, and provoked by a specific word or phrase. And there are programs that have feasted entirely on one kind of poem (e.g., haiku, sonnet, iambic pentameter) or the works of one specific poet (e.g., Homer, Milton, Shakespeare). While some of the results have passed the Turing Test when ordinary people have been asked to identify which of an assortment of poems were produced by humans (or a specific human) and which by a computer, experts (who know all of Homer, Milton, and Shakespeare by heart) have no problem in identifying poems that sound authentic but are in fact fakes.

Consider this:

*you*

*are*

*inscribed*
*in the*
*lines on the*
*ceiling*

*you*

*are*

*inscribed*
*in the*
*depths of the*
*storm* (Burgess 2016)

Or this sonnet, produced by Hafez as prompted by Marjan Ghazvininejad via "a program used to generate sonnets and other poetic forms named after the fourteenth-century Persian lyricist" (Rockmore 2020):

*People picking up electric chronic.*
*The balance like a giant tidal wave,*
*Never ever feeling supersonic,*
*Or reaching any very shallow grave.*

*An open space between awaiting speed,*
*And looking at divine velocity.*
*A faceless nation under constant need,*
*Without another curiosity.*

*Or maybe going through the wave equation.*
*An ancient engine offers no momentum,*
*About the power from an old vibration,*
*And nothing but a little bit of venom.*

*Surrounded by a sin Omega T,*
*On the other side of you and me*

which I find really quite nice.

Or this, that was actually accepted for publication in a poetry magazine on the assumption a human wrote it (Burgess 2016):

*A home transformed by the lightning*
*the balanced alcoves smother*
*this insatiable earth of a planet, Earth.*
*They attacked it with mechanical horns*
*because they love you, love, in fire and wind.*
*You say, what is the time waiting for in its spring?*
*I tell you it is waiting for your branch that flows,*
*because you are a sweet-smelling diamond architecture*
*that does not know why it grows.*

And why not? How does this differ—stylistically, esthetically, philosophically—from what is recognized as poetry in most serious publications and performances today?

On the other hand, haiku are just too easy. Is this real, or Memorex?

*Meandering thoughts*
*A sparkling lake connecting*
*after the talking*

This haiku was generated after the program learned the rules of haiku, ingested lots of words, and responded to the prompt, "lake" (Strineholm 2021).

Some may argue that these programs have not yet mastered the essence of what distinguishes poetry from schlock. But they do sometimes come up with startling and thought-provoking phrases that seem to rival what humans call creativity:

*on a charred spinning wheel,*
*the world was cold the soul of the storm,*
*the shadow's soul where the strong she still,*
*the stars that beautiful and strain,*
*and the strange and the storm of the stars,*
*and the stars of the storms of the stars,*
*i say i shall be the made the stars of the storm,*
*the stars when the wind of the stream of the shadow,*
*the thing of the said the world was a sea,*
*and the shadow of the sky* (Ilić and Moro 2018: 162).

However my prize for the most creative AI-generated phrase goes to Ghazvininejad's Hafez that "spat out one sonnet that included the phrase 'Honky Tonkin Resolution.' According to (Kevin) Knight, the phrase appeared only once or twice in Google search results: Hafez was quite literally creating original work—and maybe was a child of the sixties" (Rockmore 2020).

My favorite, oft-recited poem about computers and the future (admittedly written by a human) is by Richard Brautigan, "Machines of Loving Grace". He may have meant it as satire but I take it as vision. Except for now-nostalgic images of computers, it is still a vision to which I aspire and in some ways participate now. Stanza one is an image of mammals and computers "in mutually programming harmony". Stanza two has deer strolling "peacefully past computers". The third and final stanza ends with a dream of a future:

where we are free of our labors
and joined back to nature,
returned to our mammal
brothers and sisters,
and all watched over
by machines of loving grace (Brautigan 1968).

I couldn't have said it better myself, except we are not "returned" "back to nature" but rather evolve as posthuman nanobiotechnologies of "loving grace."

Long ago, David Miller, then a professor at the International Space University in Strasbourg, France (and the Oklahoma State University, not in France), taught me that AI is a constantly moving target, always in the future, never achieved in spite of truly amazing things that toasters, watches, automobiles, airplanes and scores of other machines have in fact learned to do. Professor Miller taught me that "Artificial Intelligence is what machines can't do yet." Once a machine can do it, then it just becomes an unremarkable aspect of our environment, and we focus on other things machines can't do yet, but that humans can do, perhaps in order to hold fast to the belief that humans are still the masters of terrestrial intelligence.

I later learned that Miller's dicta may have been based consciously or unconsciously on the following statement by Larry Tesler: "Tesler's Theorem (ca. 1970). My formulation of what others have since called the AI Effect. As commonly quoted: Artificial Intelligence is whatever hasn't been done yet. What I actually said was:

Intelligence is whatever machines haven't done yet. Many people define humanity partly by our allegedly unique intelligence. Whatever a machine—or an animal—can do must (those people say) be something other than intelligence" (Tesler nd).

Aye, there's the rub. People insist on holding on to the myth that humans are the crown of creation; "a little lower than the angels"; "nature thinking back on herself", whereas we are merely one tiny ephemeral phase in a process of infinite metamorphosis.

## 11.2 So, Why Humans?

Some years ago from a source long since lost, S. Seigal asked "Why did DNA make humans?" His answer, plus a few words (not in italics) I have added at the end, was:

> *Cells, as we know them, can be looked upon as inventions of nucleic acids to provide themselves with a local environment optimally suited to provide the materials and conditions required for nucleic acid replication.*
>
> *Similarly, the evolution to multi-cellular animals and plants can be interpreted as devices evolved to permit DNA to exploit all terrestrial space, including the land, the seas, and the air.*

*Until the last few years, one might well have wondered why DNA invented humans. It is now evident that humans were invented to provide DNA with a vehicle* that could invent artificial intelligences, artificial lives, and artificial environments, including those for elsewhere in the cosmos where humans are not fit to go.

Kim Parko ends her poem, "Encarnation," this way:

The goddess circles like a sea eagle
and loves each life
with equanimity.

The way it finds a niche.
The way it finds a host.

She loves
the rock and the water.
The crystalline veins
and fluid branches.

She loves the galaxies spiraling detritus
arms, tossing out a sphere
here and there
to be ravenously
lived upon (Parko 2020).

This disconcertingly-titled poem with its startling last stanza immediately reminded me of the Gaia Hypothesis, Goddess Theory, Earth as a brood ball for human dung beetles or the yolk sac for the fertilized embryos of broody hens, and Buckminster Fuller who wrote, "My image of humanity today finds us just about to step out from amongst the pieces of our just one-second-ago broken eggshell. Our innocent, trial-and-error-sustaining nutriment is exhausted. We are faced with an entirely new relationship to the universe. We are going to have to spread our wings of intellect and fly, or we will perish" (Fuller 1969).

All of history documents that humans have "most ravenously lived upon" the Earth. Humans are clearly planet eaters. And that has been bewailed and condemned. But maybe this is exactly what Mother, "who loves each life with equanimity", intends, as painful and precarious as this may be. We have just about devoured this brood ball but She has tossed out many other spheres here and there for her children to go and feast upon.

Now of course I don't believe that DNA or any purpose drives, or any destination lures, evolution. Evolution is not teleological. It just is. Change perpetually happens and you either adapt to it or you don't. It is generally not a good strategy to hold on to what worked in the past when the environment and its participants change rapidly—if you want to survive. Extinction is a clear option too. It is better to die with dignity than to continue to limp along in obsolescence.

Indeed it is clear that many *homosapiens, sapiens* are working hard now to create *homo cyberneticus, homo hybridus, homo machinus,* and *robotus multitudinous.* I urge them to work faster and better so they can take over before climate change and other features of the Anthropocene Epoch screw up humans and Earth beyond transformation.

The British futurist, Ian Pearson, developed the following diagram (Fig. 11.1):

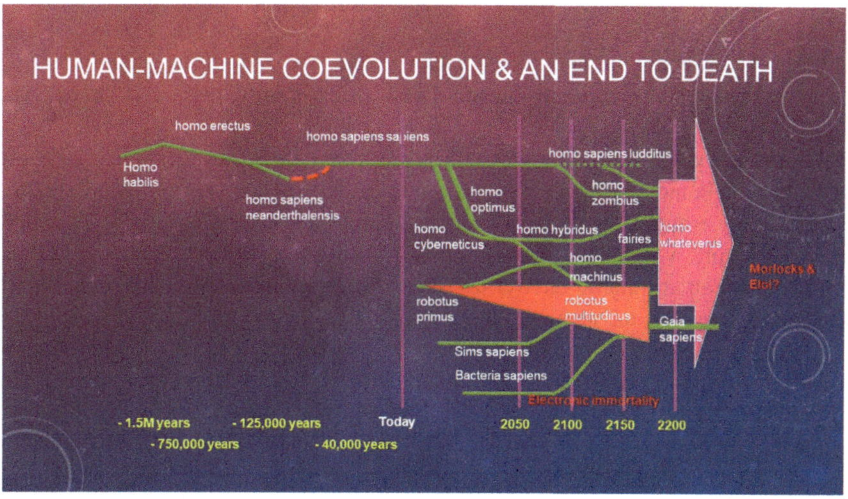


**Fig. 11.1** Human–machine coevolution and an end to death. Used with permission of Ian Pearson

This figure not only outlines plausible fuures for humans based on different foundations but also reminds us that for most of human history, *homosapiens* were not alone. There were other *homo* species among us with whom we cooperated and competed—*Neanderthal, Denisova, Floresiensis,* probably others. *Homosapiens, sapiens'* apparent monopoly on intelligence has only been for a very brief period of time, and I believe it is coming to an end.

In a sense, artilects may be returning us to what is normal in evolution—diversity of response—while ending the brief abnormal period of human's monopoly. We may be entering a period of transition from *homosapiens, sapiens*, to *homosapiens, multitudinous.* Some largely biological. Some largely mechanical. Some viral. Most a combination of alll that was and whatever comes next—ET, perhaps. These new species of *homo* should be better able than humans are to thrive in a dying Earth as well as in the vastness of extra-terrestrial space.

But wait. Did humans ever actually have a monopoly on intelligence on Earth or was that notion just evidence of how profound our ignorance and hubris is?

## 11.3 Intelligent Animals, Plants, Trees, Microbes, Fungi...

Once upon a time most humans believed everything on the Earth—and Gaia, the Earth itself—was alive, sentient, and could and should be communicated with, prayed to perhaps, asking for help and mercy. Most humans felt a special affinity to some specific animal, plant, tree, rock, mountain, valley, body of water. In Hawaii each family has an *'aumakua*—that might be understood as a protective god who usually also takes the form of some animal, plant, rock.... The 'aumakua of the family that sponsored the Healani canoe club that I once paddled for was a shark, and woe be anyone who said bad things about sharks. Their presence around our canoe was a blessing and comfort, and not a threat or worse. I have ridden a series of Honda motorcycles over my life and named each of them Aiko. More importantly, I never failed to thank them for taking me safely out and returning me safely home. And they have never failed to do so.

Almost 50 years ago, when primitive computer instruction software was being developed, I demonstrated a computer-based arithmetic lesson (which I did not design!) to a group of elementary school children during a science fair. A simple animated figure would show a smiley face with the words, "yes, you are right!" if a student got an arithmetic answer correct, or a frown and the words, "no, you are wrong. Try again" if the answer was wrong.

One bright little girl got a long string of smiley faces. I wanted to show her and the other students what the computer would do if she got a wrong answer. But she did not want to give a wrong answer. She only wanted to give correct answers.

After much persuasion on my part, she finally agreed very reluctantly to give a wrong answer. When the computer showed a frowning face and the words, "no, that is wrong", the girl burst into tears and cried inconsolably—"I know, I know. I know the right answer. He made me say the wrong answer." She turned to me: "Tell him I knew the right answer and that you made me tell the wrong answer. Tell him so he will not think that I am dumb."

I told the computer that the girl knew the right answer and gave the wrong one on my orders, but the girl was not happy because the computer was not able to acknowledge that properly. She was very peeved with me and ashamed that the computer thought she had made a mistake when she had not. She cared how the computer judged her.

Early science declared such beliefs to be myths or superstitions, and maybe they were, but recent scientific research makes it very clear that many, and probably all, animals exhibit intelligence, memory and recall, problem-solving, unselfish sharing, sacrifice, tool-making and using, creativity, curiosity, joking, deception, deceit, misdirection, embarrassment, regret, grief, love—well, essentially all behaviors and emotions that humans exhibit—but not necessarily in the way humans do. I always considered it rather silly—indeed, cruel—to try to make simians learn to speak English given their lack of the physical mechanisms that humans have for speech that other primates do not possess. They may be intelligent enough to learn to speak English but they may also be intelligent enough not to bother to try, unless pressed to please their trainers and breadwinners. Many early tests of animal intelligence expected animals to exhibit what humans call intelligence, rather than humans exhibiting enough intelligence to learn to think like other animals in order to discern actual animal intelligence. Almost every day, there is a report of some new evidence of animal intelligence even though I suspect a lot of actual animal intelligence is beyond our ability to ken. We may be interpreting certain animal behavior as exhibiting behavior we associate with intelligence in humans, but it may mean something entirely different to the animals themselves.

Once upon a time, singing and talking to plants or petting their leaves to encourage them to grow or thank them for doing so was viewed as batty. It may be batty, but we are also learning, almost every day, that plants do seem to exhibit intelligence and concern for the welfare of other plants, while the apparent cunning and care of trees for each other is stunningly impressive. They not only communicate with each other, but it seems they share resources and warn one another of dangers from trees that are out to get them. In certain situations, trees may indeed favor their own kind and lineage over strangers, and attempt to eliminate competitors.

Frantisek Baluska and Stefano Mancuso point out "that over the past decades, plant science has revealed that higher plants are much more than just passive carbon-fixing entities. They possess a plant-specific intelligence, with which they manipulate both their abiotic and biotic environment, including climate patterns and whole ecosystems. Considering plants as active and intelligent agents has therefore profound consequences not just for future climate scenarios but also for understanding mankind's role and position within the Earth's biosphere" (Baluska and Mancuso 2020: 1).

Though largely ignored until recently (and even now), Baluska and Manusco remind us that "Albert Seward in his book *Plants: What they are, What they do* (1932) was among the first to speculate that plants might in fact be superior to many animal species…. From the plant's perspective, domestication is not just subservience but rather co-evolution whereby both partners benefit from each other. Crops and many medical plants produce chemicals that alter human brain chemistry, physiology and behaviour, similar to flowering plants that manipulate insects to become their pollinators and bodyguards. Our tight co-evolution and the reliance of humans on plants to prove food, medicines and recreational drugs might raise the question of who actually domesticated whom" (Baluska and Manusco 2020: 23). So, it may not be that agriculture was humanity's biggest mistake. Rather, humans were duped by plants into doing all the hard work, freeing plants to bask lazily in the sunshine and breeze.

Research published in the *Proceedings of the National Academy of Sciences*, as reported by Diana Lutz, showed that microbes and fungi engage in what researchers interpret as a kind of rational choice theory that some economists use to study human economic behavior. "Single-celled organisms had been shown to avoid bad trading partners, build local business ties, diversify or specialize in a particular commodity, save for a rainy day, eliminate the competition and otherwise behave in ways that seem to follow market-based principles" Lutz writes. A report by Toby Kiers and her colleagues found that "fungi compare the resources on offer by different plants, and adjust their resource allocations accordingly. Some fungi even hoard resources until they get a better deal. 'We now see that such playing of the market happens in microbes. Microbial traders can be ruthless, even using chemicals to actively elbow competitors out of the marketplace,'" Kiers concluded according to Lutz (2014).

Recent research on fungi shows them to play a vastly more important role in sustaining and regulating life on the planet than realized before. Fungi are classified by scientists into their own kingdom, separate from plants and animals. And yet the last thing fungi are is "separate" from plant and animal life. They are the very basis of life, and almost entirely ignored or misunderstood.

Zoe Schlanger states that fungi "form large networks of *hyphae* strands in order to feed. These strands, when massed together, are called *mycelium*. The total length of *mycelium* threaded through the globe's uppermost four inches of soil is believed to be enough to span half the width of our galaxy" (Schlanger 2021). The largest and oldest single living organism may be an *Armillaria ostoyae* fungus in Oregon (Casselman 2007). It is thought to be between 1900 and 8650 years old and weighs somewhere between 7,500 and 35,000 US tons. It is the biggest of four other *Armillaria ostoyae* in the same area all of which are almost entirely hidden underground. "Yet despite fungi's pervasive presence in the natural world, scientists estimate that only 6% of fungal species have been described. (There are believed to be between 2.2 and 3.8 million of them in existence—six times the number of plant species.)" (Schlanger 2021: 3).

Fungi are also a threat to life. Rose Eveleth observes that "Humans should declare ourselves lucky that they don't have to constantly worry about fungal infections. 'If you were a tree, you'd be terrified of fungi,' says Dr. Arturo Casadevall, a microbiologist at Johns Hopkins university who studies fungal diseases. And if you happened to be a fish, a reptile, or an amphibian, fungus would also be quite high on your list of fears... (Fungal infections are known to wipe out snakes, fish, corals, insects, and more.) In recent years, a fungal infection called *Batrachochytrium dendrobatidis* (chytrid) has decimated amphibian populations around the world, with some scientists estimating that chytrid is responsible for population decline in over 500 amphibian species. To put that into context, that's around one out of every 16 amphibian species known to science."

> Fungi are more closely related to humans than they are to viruses or bacteria, which means that, in general, things that kill them also kill us.... Already, over 300 million people globally contract serious fungal infection each year and over 1.5 million of them die. So what happens next, and what should we do? Casadevall gave the same answer every scientist gives to this question: The field needs more funding. 'Humanity should be investing more in learning about what is the largest kingdom on the planet,' he says (Eveleth 2021).

## 11.4 Past and Next Steps

Consider the evolution of reproduction, from *accretion, three and a half to four billion years ago*, to the present and near future:

1. *Accretion (3.5–4 billion years ago)*

   Fission: Isomorphic replacement, Crystals,

   Blue-Green Algae;

2. *Replication (2 billion years ago)*

   Fission: Single cell division,

   Amoebae;

3. *Bisexual gene recombination (1 billion years ago)*

   Fusion: mutual growth/generational differences/mutations/perpetual change:

   the evolution of plants, animals, humans;

4. *Prosthetic as well as biological enhancement*

   Clothes, houses, eye glasses, shoes, artificial limbs

   Cellular/Organ *transplants*

   Cellular/Organ *regeneration*

   *Synthetic* cells/organs;

5. *Genetic engineering*

   200 KYA—Marriage and Incest rules

   5–10 KYA—Agriculture/Animal Husbandry

150 YA—Hybrid selection

50 YA—Augmented animals, Dolly

*Soon*:

Clones

Chimeras

*Transhumans*

*Posthumans*

6. *Artificial Life and Intelligence*

Electronics

Computers

Internet

Artificial life

*Mobile, sensing, responsive, independent*

artificial intelligence;

Many varieties of post-*homosapiens* (Inspired by Lock Land 1973).

One important thing to notice is that over the last 15,000 years, most of the new modes of reproduction have been artificial—the result of human actions, intentional and unintentional. Humans have modified ourselves and our environment without serious restraint for most of our existence. Everything once "natural" has already become substantially "artificial." Whatever happens from now on will result even more from human activities.

We have different technologies now, but the impulse to change and improve nature is fundamental to humans. We must recognize that, and take responsibility for what we have done and are continuing to do by consciously striving to Govern Evolution.

Moreover, primarily but not solely because of human action, the planet into which humans evolved in the Holocene Epoch is not the planet we live on now. Many more changes will occur to Earth during and after the present Anthropocene Epoch.

As Stewart Brand said, "We are as gods and might as well get good at it."

That attitude may be scientific hubris and domineering patriarchy at their worst. Or this may be humanity at its nurturing, sacrificial, feminine best. Or beyond any analogy based on the meaning of "gender" now. Moreover, we can't avoid the consequences of our actions. So we must do our best to surf the tsunami from now on.

To argue for the necessity of participatively designing and achieving a world of artificial intelligence, artificial life, synthetic and synthesized environments, artificial "nature" and the rest may seem ridiculously at odds with the necessity and urgency of addressing the multiple overwhelming novel complications of global climate change and unemployment given the enormities of the difficulties to surviving climate change.

Perhaps. But to the contrary it may be that it is in league with our emerging artilects that the futures of humanity depends. Artilects might acquire the wisdom, diligence, and foresight necessary to help humans surf the rising tsunami of change. Humans don't seem capable of doing so on our own.

## References


Baluska, Frantisek, and Stefano Mancuso. 2020. Plants, climate and humans. *EMBO Reports* 21: e50109.

Brautigan, Richard. 1968. Machines of loving grace. In *The pill versus the Springhill mine disaster*. Four Seasons Foundation.

Brockman, John, ed. 2019. *Possible minds: Twenty-five ways of looking at AI*. New York: Penguin Press.

Burgess, Matt. 2016. Google's poetry was written by an AI system after it was fed thousands of unpublished romantic novels. *Wired*, May 16.

Casselman, Anne. 2007. Strange but true: The largest organism on Earth is a fungus. *Scientific American*, October 4.

Dator, Jim. 1980. EIES and Racter and me: Computer conferencing from a Pacific Island. In *Pacific Telecommunications Conference*, ed. Dan Wedemeyer, 3A-1–3A-10. Honolulu: Pacific Telecommunications Council.

Dator, Jim. 2019. *Jim Dator: A noticer in time. Selected work, 1967–2018*, 154–156. Springer Nature Press.

Eveleth, Rose. 2021. It's time to fear the fungi. *Wired*, November 23.

Fuller, R. Buckminster. 1969. *Operating manual for Spaceship Earth*. Southern Illinois University Press.

Ilić, Suzana, and Martina Jole Moro. 2018. Multimedia art: The synthesis of machine-generated poetry and virtual landscapes. In *Art machines: International symposium on computation media art proceedings*, ed. Richard Allen. School of Creative Media, City University of Hong Kong.

Lee, Kai-Fi. 2021. Why computers don't need to match human intelligence. *Wired*, December 16.

Lock Land, George. 1973. *Grow or die: The unifying principle of transformation*. Random House.

Lutz, Diana. 2014. https://source.wustl.edu/2014/01/microbes-buy-low-and-sell-high/.

Parko, Kim. 2020. Encarnation. *Poetry*, 220/2, May.

Rockmore, Dan. 2020. What happens when machines learn poetry. *New Yorker*, January 7. https://www.newyorker.com/culture/annals-of-inquiry/the-mechanical-muse.

Schlanger, Zoe. 2021. Our silent partners. *New York Review of Books*, October 7.

Scott, Andrew, José R. Solórzano, Jonathan D. Moyer, and Barry B. Hughes. 2022. The future of artificial intelligence. *International Journal of Artificial Intelligence and Machine Learning* 2 (1).

Strineholm, Philippe. 2021. Exploring human-robot interaction through explainable AI poetry generation. Thesis for the degree of Master of Science in Engineering—Robotics, Malardalen University School of Innovation, Design and Engineering, Vasteras, Sweden.

Tesler, L. (nd). http://nomodes.com/Larry_Tesler_Consulting/Adages_and_Coinages.html.

Turoff, Murray, and Starr Roxanne Hiltz. 1978. *Network nation: Human communication via computer*. Reading, MASS: Addison-Wesley Publishing Company.

# Chapter 12
# Technology, Values and Change

**Abstract** A discussion of a theory of the role of technological change in social and environmental change: "We shape our tools and thereafter our tools shape us", with the current period of indivollectivity as an example.

**Keywords** Behavior · Cause · Culture · Free will · Games · Futures studies · Identity · Indivollectivity · Japan · Models and media · Progress · Reality · Technology · Time · Values

Ever since I was in high school, I have been interested in why people behave as they do, especially in the aggregate called "society" or "culture". I was interested in the big picture—why societies were constructed as they were. Why and how societies changed over time as well as resisted change—in other words, the ideas of Plato, Aristotle, St. Augustine, Joachim de Flora, St. Thomas Aquinas, Ibn Khaldun, Karl Marx, Oswald Spengler, Arnold Toynbee, W. W. Rostow and other theorists/apologists of "Development". These men were all concerned about what the purpose of life was, and how human settlements could be designed so as to enable humans to live their lives in accordance with that purpose. That impulse initially led me to political science which at the time not only emphasized the history of political theories about the state and governance, but also the use of mathematical and statistical methods for predicting the future and designing good governance fit for it.

That paragraph makes it sound that only men were influential in my intellectual life, so I need to repeat my debt to my mother, aunt, grandmother, and to a series of female and trans teachers, scholars, administrators, colleagues, pupils, spouses, and children who were equally, if not much more, influential in how and what I think and feel, believe and act.

But it was a specific awakening in Japan that drove me fully into futures studies. John Randolph asked me to review a paper of his that used the theories of Oswald Spengler to understand the history and future of Japan. His conclusion was that Japan indeed went through the same stages that Spengler described for the west, in the same sequence, and each for approximately the same length of time, but that in every stage, Japan was 200 years ahead of the west (Randolph 1964). I was stunned speechless. Ahead of the west? How was that possible in 1964? I wondered. But more importantly, what does this mean for the west? Should we look to Japan for

 
J. Dator, *Beyond Identities: Human Becomings in Weirding Worlds*,
Anticipation Science 7, https://doi.org/10.1007/978-3-031-11732-9_12

our future, just as we taught developing countries to look to developed countries for their future? What is the future of the US and the west? Development theory didn't say. Nations were undeveloped, developing, or developed, and that was it. No! I said, that can't be it. First of all, that pseudo-evolutionary way of telling the story of "progress" is very narrowsighted, if not false. To the extent it is true, or at least useful, then there must be something beyond development, and I wanted to know what it was, or could be, or should be, and so I flung into the company of those who invented futures studies.

Futures studies is grounded in various key concepts such as time, cause, reality, free will, where "the future" is "located" and the like. Of course, futurists often disagree on the details of each of these concepts.

Does time objectively exist or is it a human construct derived from features of our biology? Is it not so much that "time passes" as that what we call "time" are the changes we experience as we age? Or is time a fundamental feature of the cosmos, though each of us is captive to our local experience of it? "Time is what keeps everything from happening all at once", and yet on the cosmic scale everything *is* happening all at once. It is only our limited biological lives on this minor planet in this obscure corner of the universe that misleads us into believing that time exists and somehow "passes". Time wasn't really important (or a concept) until it could be measured and commodified.

What actually is "cause"? And since it seems that everything is caused by everything else, does it make sense to talk about cause at all? While fathers and mothers cause children, children also cause fathers and mothers.

What is reality? Each one of us perceives a significantly different and ever-changing world. A kind of agreement is forced on us by our native language, but each language asks us to share a slightly (and sometimes significantly) different world from other languages.

Does that falling tree in the forest actually make a sound when no one is there to see and hear it fall? If it falls at all. If it—or the world itself—exists at all apart from humans' observation. Is Heisenberg only going on about waves and particles or about everything?

Is "free will" just another delusion? Accumulating evidence suggests that "we" act and then we decide to choose to act. "Only the fool, fixed in his folly, believes he can turn the wheel on which he turns", a cat-loving poet declared. And yet the same research suggests that the brain is constantly forecasting the immediate future and learning from and improving forecasts based on feedback from experience. Futures orientation should be a prime feature of human behavior, but it most certainly is not. And yet as I keep saying the only arena of time over which humans might have some purposeful influence (if there is any at all) would have to be the future.

Does the future lie "ahead" as commencement speakers claim? Do we "face the future"? If so, why can't we see it more clearly? Or is the future behind us while we face the past which we do see clearly, though less and less so as it fades over the horizon? Is the reason we are often surprised by the future because it arrives from in back of us, as traditional Hawaiians (and others) have it?

Modern futures studies which began gaining traction in the early 1960s was a worldwide phenomenon with various strands. One was dominated by English speakers (though most were not native English speakers, they were educated to some extent in English and used it for many of their discussions of the field) who not only apparently accepted unquestionably all the memes built into English, but also all the beliefs about reason and progress and development and time and cause and free will built into them by formal western education since the Enlightenment, the Protestant Reformation, and Newtonian Physics. These were (for the most part) unquestioned bedrocks of reality upon which all squabbles and battles among westerners were based.

Ideas from the Enlightenment onward were problematized by some futurists while accepted unthinkingly by others. For every futurist supporting better foresight for development there was another opposed, hugging trees and favoring the preservation of "natural" environments over their destruction. Most futurists would agree that rational decision-making is preferable to reliance on past myths and superstitions, but is rationality possible for humans? Doesn't the evidence from psychology, psychiatry, sociology, brain science make it clear that we do not think carefully before we act, even when we think we are thinking carefully. Our emotions, prejudices, distractions, false memories, short little spans of attention that keep running down the alleyways make it impossible for anyone to be rational in the sense contemporary law and economics claim we are or can be.

All humans evaluate things; have preferences, fears, hopes, beliefs about what is good, bad, holy, secular, desirable, undesirable, right, wrong, essential, frivolous. While many people share values with members of their families, communities, nations, cultures, and perhaps all humanity, people often hold and act on the basis of some values that are different from their closest companions as well as their most distant enemies. There may be some people who have no values at all—who make no distinction between the words I listed above—but they are few, I observe.

I believe values and their acquisition are similar to language and its acquisition. There may be some humans who do not speak for various reasons, but most humans do learn to talk, and they learn to speak whatever language is spoken around them, with individual variations. Learning one's "mother tongue" is duck soup for most of us. It is impossible for most of us not to learn it. Learning many languages is also easy when you are young and are also exposed to them regularly, but very difficult for most people to learn later, either at all, well, or without "an accent".

The sounds that humans can produce are vast. Babies in their earliest stage of babbling may utter them all, but they soon learn to babble and then to talk using only the far more restricted number of sounds that carry meaning in their mother tongue. Meaningfully different sounds in one language may be almost unrecognizable in other languages. Learning to hear and articulate the differences may be a mark of humor or ridicule of nonnative speakers. When my Japanese students told me that they prayed all vacation, I was struck by their devotion until they also told me they rubbed flied lice. The sounds signified by *th* in English are among the last that even native speakers learn, and many speakers of other languages may never master. It is far too easy to ridicule many Americans when they try to pronounce words in anything but the variety of English that they were born into.

So also with values. The range of values is vast as well. However, each person, each culture, has a limited number of values that children learn effortlessly and early, though the values one learns in one family or community might be quite different from those learned in others.

But what is crucial about values is not the words used to label them but the actual behavior that is permitted or proscribed.

Our behavior strongly influences what our true values are. Many values we say we hold struggle to influence our behavior. We do not have values telling us to do or to not do things that we can't do. Our values only address things we can do, and often do, whether our values say we should or should not. Behavior is determined by many things—genes, physiology, the environment, culture, technology, education, experience, and old values. But those old values are themselves based on past behaviors.

While genes, physiology, the environment, culture, experience, education and especially old values do their best to regulate our behavior today, new technologies have become the major source of social and environmental change. New technologies make it easy or indeed necessary to do something that was difficult or impossible to do before, and render difficult or impossible things that were once necessary and perhaps desirable.

Old values are based in part on behavior enabled by old technologies. New technologies enable new behavior that encourage new values based on the new behavior and thus produce social conflict and strife, continuously destroying old social systems and enabling the creation of new ones. It is not terrorists, or communists, or radicals who are the true revolutionaries in our world. It is scientists who imagine new technologies, engineers who create them, entrepreneurs and business people and advertisers who create demand for them, financiers who invent debt instruments that let people "buy" what they could not possibly acquire otherwise and so change their behavior and values as they use them.

In *Mutative Media,* John Sweeney, Aubrey Yee and myself show how the evolution of modes of communication have changed the ways humans think and act (Dator et al. 2014). We start with the evolution of speech—probably the most profoundly powerful technology of humans—through writing (by hand), through the printing press (in Asia and Europe), and then through the stunning array of audio-visual communication technologies enabled by electricity and electronics (radio, TV, computer, Internet, social media) and imaging (analog—still photographs, movies, television—and digital), as these technologies merged and altered time, space, distance, and life.

We adopted a broad definition of "technology" as "how humans get things done" rather than pieces of physical hardware alone.

We discussed *three views of technology*—that it is *neutral* (neither good nor bad, but up to the user); that technology is *demonic* (driving us farther and farther from the Garden of Eden and our true humanity while also destroying the Earth which is the very basis of our lives); that it is *mutative*—it is not neutral since it inevitably changes our behavior which challenges our values. For years I had called this third view of technology as "transformative", but John Sweeney convinced me to understand that "transformation" is one positive view of the mutation. Others may view the mutation negatively. We understand mutation in technology to be similar to the process by which evolution has driven all of the diverse forms of life and living since life began on Earth, the most important aspect of that evolutionary process being that it is purposeless, goalless, not "progress" but simply a change of form and function that either proves to be evolvable, or not, in which case it may die out.

I need to thank Jan Huston for making clear to me the profound limits of the notions of "sustainability". Evolution is not "stable"; it is dynamic, uncertain, destructive and constructive. Forms and processes that we favor need to be evolvable, not sustainable. They should enable the mutation itself to mutate and evolve.

There are *three kinds of technology*. *Physical*, the tool or method itself which is typically what people mean when they refer to technology. But there are two other kinds of technology—biological and social. *Biological* technologies are those ways of doing things that evolved through biological evolution—breathing, eating, sleeping, defecating, walking, sexual intercourse. They are—or were—such dominant ways by which humans "got things done" from the beginning that they are often not recognized to be technologies until they are threatened or replaced by social or more likely physical technologies. The third kind of technology is *social*—usually some institutional way of "getting things done". Schools are technologies for education and socialization (and other things). Babies may be produced by biological technologies—sexual intercourse—but once born, they are protected, fed, nurtured by social technologies such as the family.

The *three aspects of technology* (hardware, software and orgware) are the most likely to be misunderstood or ignored. Hardware is mistakenly equated with physical technology as the sole aspect. But as we have learned from computers, every hardware needs a software to make the hardware useful. In the absence of the invention of writing and learning the rules of writing, a pen is just a hair pin, a single chopstick, a weapon. It is not a writing instrument until someone invents, learns, and teaches the software of the rules of writing. And there can be no writing instruments, or rules, or paper, or newspapers, or blogs until the *orgware* for the hardware and software is invented and functioning—the human organizations like corporations, labor unions, laborers and managers, and the like. This is where most social conflict actually manifest itself—when the hardware, software and especially orgware for a new technology takes the jobs of the orgware of the obsolete technology.

And it is this replacement that is happening big time in the world for the last 200 years, ever-accelerating.

Finally, technology almost never causes change instantly and across all scale. There are *six phases in the lifecycle* of all technologies: *invention, development, diffusion, maturity, obsolescence, and death* (and/or *resurrection*). Significant social and environmental change does not happen until the diffusion stage. The initial invention and even development of a technology seldom has any significant immediate impact, and the depth and length of the impact depends on the extent and spread of the diffusion which is seldom smooth and steady. There typically are early adopters, mature users, and final stragglers.

For most of human history the rate of technologically-induced social and environmental change was so slow that it was scarcely noticeable. Genes, physiology, the environment, and culture were dominant. But we live now in a period of extremely rapid technological invention, development, and diffusion—and of planned obsolescence. We have barely learned how to use the latest version of Microsoft Office before a new version comes out, with buttons that had been on the top left spread across the keyboard, but mainly on the bottom right. Productivity decreases and errors and frustration increase without end. According to Arthur Levine when he asked a student how she adapted so easily to Google, Yahoo, cellphones, and Skype, she replied "it's only technology if it happened after you were born" (Levine 2012). The technologies already widely used when one is born, or shortly after, are virtually invisible to that generation. They are the water in which, as fish, they swim. However that stream is flowing very quickly and one isn't a digital native very long before they are a fumbling old fuddy duddy. Skills that used to last a lifetime—or countless generations—are obsolete by the time one is 30.

For example:

## 12.1 Indivollectivity Now?

For most of human history, individuals and communities have lived in "one present" and looked forward to "one future", defined by one set of technologies. For most of human history, technological change was rare, slow, and the social and environmental consequences rarely noticed. During most of this time, people lived and thought collectively in small face-to-face groups, not separately and individually. They had a sense of self and other, and therefore some kinds of individual agency. But there was no concept of "privacy" or "my individual rights". Some Greek and Roman philosophers eventually conceived of those ideas, but the overwhelming majority of humans neither imagined nor experienced privacy or individual moral freedom in their daily lives. Indeed, the social value and impact of the Greek and Roman philosophers was near zero until they were rediscovered in the late middle ages. Community- and family-focused life dominated for most people until the scientific-industrial revolution and events leading up to it (the Reformation, Renaissance, and Enlightenment) when more people began using technologies that gave them first the experience and then the idea of individualism and individual freedom while rediscovering/inventing texts that explained and justified their experiences.

From the middle of the seventeenth century until the middle of the twentieth century, modern societies were increasingly dominated by individualizing technologies, spurred by the printing press and culminating with the automobile and the telephone. During that period, each new technology seemed to free the individual from the confining traditions and bonds of the community, eventually creating seemingly autonomous individuals each with their own unique sense of self and their future—their personal values and beliefs—(give me liberty or give me death; it's my way or the highway) leading to the opportunities, ideologies, triumphs, and catastrophes characteristic of modern times.

From the mid 1950s, however, new technologies emerged that tended once again to collectivize, though often at a global level and certainly in conflict with values and institutions based on earlier, local, collectivizing technologies. The first of these new collectivizing technologies was television (creating what Marshall McLuhan called "the global village"); the second was the personal computer when global networking began and expert authority died; and the third are the currently popular social media and the emerging hive mind (or, Noosphere, as de Chardin anticipatorily named it).

Something else unique also began to occur during early modernity, still accelerating today: technological and social change became so rapid that individuals and communities were caught for the first time in a whirlpool of conflicting technologies, values, and institutions, some of which were obsolete and vanishing, some were old and fading, some were current and thriving, others were new and emerging, and others still vividly imagined but not yet achieved. Until a 150 or so years ago, everyone in a community used and were influenced by the same technologies. Not now.

At the present time, oldest age-cohorts live by vanishing and fading technologies, values, and institutions; middle age-cohorts by fading and thriving technologies, values, and institutions; while younger cohorts embrace emerging and imagined technologies. Nonetheless, each cohort, individual, and the community as a whole is possessed in some measure by all levels. This is new to human experience—each age-cohort living in substantially different worlds but at the same time and place with other cohorts. Intergenerational communication and easy understanding is difficult. Cultural chaos reigns in every part of the world.

Agricultural—indeed, pastoral—era metaphors, institutions, and values still persist. With the printing press as the iconic technology and, in the West, The Bible that the printing press liberated from the Church as the iconic text, we still say God is a king, with heaven ("above") and Earth ("below") as his kingdom while we are his subjects with no rights or wills of our own. God is our Shepherd and we are His ignorant and willful sheep in need of correction and protection by his rod and staff. Similar nomadic and/or agricultural myths and metaphors exist in every culture in the world even when the memories much less the experiences of sheep and shepherds have long faded away.

Another very clear example of how the rhythms derived from agricultural societies still control us today is found in many academic and legislative calendars and holidays worldwide. Both learning and legislating are still often part-time activities, originally scheduled in the Northern Hemisphere for the late fall and winter when the crops were in, to allow their participants to recess from school and go home to help plant and reap from late spring to early fall. This once-sensible practice now makes no sense at all in the industrial or informational world, and yet seems impossible to change until all agricultural experiences and images finally die. Even though we live in a global world in many ways, we still retain local loyalties in sports, and admire, if not emulate, values of small communities depicted in TV shows and advertisements. Now, with rise of and response to Trump and Trumpists as well as other hypernationalists around the world, tribalism is challenging globalism as the dominant perspective once again.

For the most part, transportation technologies defined and dominated the industrial era, producing first the railroad, then the automobile, then the airplane, creating both the city with its suburbs and the (often continental) nation as its iconic institutions. The automobile, for the first time enabling true auto-mobility, was the iconic technology. Nothing provoked the sense of individualism, freedom, personal identity, and social irresponsibility more than the automobile. It is only a slight exaggeration to say that the fall of communism began when private individuals in communist countries were allowed to own automobiles instead of keeping them herded on mass transit.

The allure and power of the automobile is still extraordinary. Everywhere it spreads throughout the world, it transforms stable, obedient, community-focused peasants and workers into roaming, pleasure-seeking, death-provoking adolescents of all ages. Commuting, "rush hour" traffic jams, horrendous deaths and injuries, environmental pollution, and oil wars are its side-effects, in spite of which the attraction of personal identity through sports car/SUV automobility is far too strong to allow telework to end commuting, or for other forms of transport, especially bicycles and walking, to end pollution and oil wars. The emergence of self-driving cars can only spur free-spirited individuality—until the oceans rise and the oil runs out.

Responses to Covid-19 may have weakened—perhaps ended—resistance to teleworking and many young people are eschewing automobiles. But they are not clamoring for mass transit. They adopt even more individualizing modes of transport—skates, skateboards, bicycles, unicycles, scooters. Some still prefer the ones that require human power to propel them, but most, it seems, prefer them motorized with all the accompanying danger, congestion, clutter of automobiles as rampaging wolves in packs.

As individualizing transportation technologies shaped industrial societies, so also do collectivizing communication technologies define information societies. First movies and then television were the initial icons, followed by personal computers and the Internet, and now social media. Major social consequences of these technologies include the focus of life becoming advertising-fueled "shop till you drop", along with the mania of tirelessly working 24/7 at meaningless networked jobs in order to impress your fellow workers and your boss; the end of the human expert/authority and the rise of personal and peer facts and fantasies; and the dominance of entertainment, "fake news", games, professional sports, and of virtuality over "reality" in general. Wolfpacks on steroids on virtual bikes.

Demographically, more and more adults in information societies everywhere are "only children" themselves now living seemingly alone with no children, spouses, or roommates of their own, but actually interacting with myriad other humans and increasingly smart machines with clever algorithms via an ever-changing array of multitasking communication technologies, surrounded by the decaying remnants of take-away food. This novel form of individuality and community may continue to evolve as long as new electronic and molecular communication technologies continue to be produced and acquired.

The next step already underway is to lose the confining, stationary, built environment—the solitary room, apartment, house—and live in light, sturdy, mobile, self-sufficient, adaptable cocoons like the cushicle envisioned by the Archigram group many years ago: a kind of personal backpack with all of one's necessities in it that can be unfolded and joined with other cushicles whenever group interaction is sought, and then (after tidying up the common environment) folded back into one's personal backpack again as each member of the former group moves on.

The new nationalism sweeping the globe now is often seen as a revival of "blood and soil" patriotism of the old industrial days, and many people may promote and strive for that. But at the same time, the forces of individualizing-collectivizing social media seem to provoke something different and perhaps new for the immediate future—*indivollectivity* (Dator 2019).

I said at the outset of this discussion of the interrelationship between technology, values, and social and environmental change that I was drawn to political science as an academic discipline because political science at the time not only emphasized the history of political theories about the state and governance, but also was beginning to use mathematical and statistical methods to predict the future and design good governance fit for it. These were the early days of what was called "the behavioral revolution" in political science: instead of focusing only on constitutions, laws, and formal structures, as was traditional for the field, some cutting edge scholarship focused on learning from actual political behavior which required the use of rigorous methods of observation, questioning, measurement, and evaluation (especially statistics and the emerging field of computer modeling of complex human systems). So I set myself to the task of learning how to use these tools and the theories behind them.

That was not easy to do. I had been awarded a scholarship that allowed me to begin my graduate work at the University of Pennsylvania. It turned out to be very far from the cutting edge of the behavioral revolution, I soon learned. Indeed, the course I took on "American Government" was taught entirely as a course in American Constitutional Law. Everything I needed to know about American government was to be found in the decisions of cases and controversies before the US Supreme Court. Now, I had done some Constitutional Law as an undergraduate, and appreciated the "how many angels can dance on the head of a pin" medieval scholastic processes of Con Law, but surely there was more to be said about American government than what nine old men—or at least a majority of them—had to say about the few issues they allowed to be brought before them, I thought. But no. I was told that was all I needed to know since, in the US, law is what the judges say it is.

Indeed, one day, in 1955, Professor Richard Snyder from Princeton University in the far off wonderland of New Jersey was invited to speak at Penn. But only faculty members and PhD candidates of the political science department were allowed to attend. His ideas were deemed too radical for the impressionable minds of MA students such as myself. Snyder had written *Roots of Political Behavior: Introduction to Government and Politics*. (New York, American Book Company) in 1949 and I had read it, but it contained ideas deemed ridiculous by most Penn faculty members of the time. So, with MA in hand, I transferred to The American University in Washington, DC, which I knew to be a kind of hotbed of behavioralism, for my PhD. I was very, very glad I did. Later, I attended the Survey Research Center of the University of Michigan; the Second Institute on Mathematical Applications in Political Science at Southern Methodist University; and co-directed and taught in the 4th Institute on Mathematical Applications in Political Science at Virginia Tech.

Inspired by the pioneering work in judicial behavior by Glendon Schubert, I used what I had learned to research and publish about comparative judicial behavior while I was in Japan. I also conducted a large survey of attitudes of Tokyo residents that replicated a study by Gerhart Lenksi that he had titled "The Religious Factor." When I came to the University of Hawaii where the State Legislature created the Hawaii Research Center for Futures Studies, we began work on a computer model, at the urging of the President of the Hawaii State Senate, called HAFDAM (Hawaii alternative futures decision-aiding model) that legislators could use to simulate the consequences of various versions of proposed legislation before passing laws.

In other words, I took the behavioral revolution seriously for a while and did my best to learn how to think and act mathematically instead of just verbally as my mentors at Penn would have me do—but not only Penn. Earlier, when I turned in a term paper in my undergraduate Shakespeare class at Stetson University in which I came to certain conclusions about Shakespeare's soliloquies based on a statistical analysis of them, I was severely admonished by my teacher who reluctantly gave me a good grade but warned me never to do anything like that again if I wanted to prosper in academe, or at least in the humanities version of it.

But the world of a person who perceives and communicates about it in the words of common and/or formal language is quite different from the world of a person who relies on the forms and conventions of mathematics. Which way is right and which is wrong, or, rather, which way leads to truth and accuracy and which to error and vagueness?

Neither. Reality is vast—beyond our full comprehension—and each model and medium we use to apprehend and comprehend it tell us certain truths and obscures others. There indeed are "alternative facts".

As I have said, I spent the first six years of my academic career as the only nonJapanese teaching in the College of Law and Politics of Rikkyo University, in Tokyo, Japan. Since few of the students or faculty could speak or understand spoken English fluently, I did my best to conduct my classes, consultations, and research in Japanese—though in truth, their English was often far better than my Japanese. Nonetheless it was during that time that I first became personally aware of the fact that what we think we understand about the world is entirely dependent on (1) certain biological features of humans that had evolved over eons and (2) the models and media we use to perceive and communicate our perceptions of the world. The world to a monolingual native English speaker in America is, I can assure you, fundamentally different from the world of a monolingual native Japanese speaker in Japan. And while I lived in Japan, speaking Japanese as best I could, I came to see the world and act in it quite differently from the way I had previously seen the world in English in America, simply as a consequence of seeing it through Japanese grammar, syntax and vocabulary.

It was my awareness that I was a fundamentally different person in Japan from what I was in America, more than anything else, that set me off exploring the relationship between what we think with, what we think about, and how we behave.

After I left Japan, I went to Virginia Polytechnic Institute, a large school in a tiny town in the remote mountains of western Virginia. But while I was there I had the very good fortune of falling under the influence of a creative bunch of architects and artists, American and foreign, with whom I worked producing a variety of nonverbal, non-numerical models and media for political science. We developed and experimented with various prototypes for teaching university political science courses by first observing behavior and then communicating our conclusions through static, three-dimensional models, instead of relying on textbooks and spoken words.

On coming to the University of Hawaii, I designed a learning space based on what I had seen architects use at Virginia Tech where students could construct three-dimensional models with peer and faculty guidance. However, I soon was sidetracked by the opportunity to design and produce "Tune to the Future," a university political science televised course that was innovative in its use of comedy, short-takes, and multi-camera, quick-editing derived from popular shows of the era, such as *Rowan and Martin's Laugh-in* and *Monday Night Football*. We also produced a considerable amount of written support material. Many shows were live, but some were taped, and I visited all islands and met with student groups while watching and then discussing the taped segments. This course was awarded a prize for creativity from the National University Extension Association.

But I realized that I needed to know a lot more about tele-education, so I jumped at the opportunity to take a two-year leave of absence from UH, and went to Toronto where I was director of the Futures Project of the OECA (Ontario Educational Communication Authority; also known as TV Ontario). There I learned a lot from Marshall McLuhan and many others, and wrote and produced educational television and multimedia programs and support material. During that time and subsequently, I also collaborated with Simon Nicholson of the Open University in England and produced several different things in many different media for his course, "Art and Environment," one of which was a futures-oriented television program, called "Que Sera, Sera", that was shown over the BBC2 TV network for ten years.

When I returned to the University of Hawaii, I began teaching courses on media literacy through the Department of Political Science where students endeavored to gain understanding and express complex ideas audio-visually and not just through reading and writing. At one point, I conducted a semester-long series of symposia, demonstrations, and other events that culminated in a three-day "Mediacy Fair", all aimed at interesting the university and broader community in using electronic communication technologies for serious, entertaining, and effective educational purposes.

In addition to television, I also produced scores of what were once called "multi-media shows" (using multiple slide projectors, motion picture projectors, and integrated music, narration, and sound effects) for international conferences on a wide variety of topics. In addition, I also taught several courses throughout the far-flung Micronesian islands of the Pacific over the PEACESAT satellite network, sending assignments and homework back and forth over primitive facsimile machines; several "courses by newspaper"; and also several by radio.

Each one of these media made me completely rethink not only the content of my courses but also the way content was delivered. No matter how brilliant a lecturer I may have imagined myself to be, nothing was more boring than for me to stand and deliver my typical classroom/public address lecture on TV. Though infinitely more time- and labor-intensive to produce and present, the multimedia presentations were also infinitely more immersive and powerful than any lecture or book. The courses by radio and especially newspaper were something else indeed.

I learned that each medium had its own grammar, and I had to learn the grammar if I was to use the medium effectively. That is why, in spite of the success of "Tune to the Future", I decided to learn more about effective educational television production by working with TVOntario. The importance of knowing the grammar of media within specific cultural contexts was also made vividly clear to me by the book, *Through Navaho Eyes* (Worth and Adair 1972), and the process of movie-making it describes. In 1966, Anthropologists Worth and Adair sought to teach young Navaho how to make movies, using 16 mm, black and white film Bell and Howell cameras. They sought Navahos from reservations who had little or no exposure to American, Hollywood style movies, and gave them near total freedom to write their own scripts and create their own movies any way they wished. The role of the instructors was strictly technical—teaching the students how to make the cameras operate properly. But it took a great deal of restraint on the part on Worth and Adair not to constantly object to almost everything their students did—from the stories they chose to tell and how they chose to tell them, to the way they directed and filmed the action, to the way they edited the final results. None of it was done the way Worth and Adair intended them to do it, but, to their credit, they allowed the students to do it their way.

The big day during which the students would show their films to the village elders was fast approaching when one of the students who had spent a great deal of time off the reservation and seen many American movies asked and received permission to produce a second film in addition to the one he did for the assignment.

The audience reacted positively to the films as they were shown, and lauded the abilities of the youngsters to tell stories according to Navaho cultural conventions—except for the second film produced by the student which they said they did not understand since it was in English.

Now, all the movies were silent. There was no narration, sound effects, or captions. But since the movies they approved were done according to Navaho storytelling rules—and in almost complete violation of what Worth and Adair desired—while the one extra film was done more or less according to Hollywood conventions that the elders did not know or understand, they declared that it was all Greek to them.

Once upon a time, for a fleeting instance, movie-making around the world was told on the basis of local storytelling conventions. While there are still subtle (but important) differences between, say, Hollywood, Bollywood, Hong Kong, and Japanese movies, basically all movies now are told the same way (depending on the specific genre)—namely, according to American conventions that are based on the culture of capitalism: Whatever can be produced that will have the biggest box office appeal worldwide, requiring the minimum of dubbing or editing to get the story across. That means specifically a story of good *vs* evil told in brutal, grunting, shooting, chasing, violent Manichean images.

It turned out that I was in Canada, with Marshall McLuhan and company, at a kind of turning point in communication history. Given the power and popularity of cinema and television, and the theories of McLuhan, it seemed clear that "The Word Was Out"—that when TV entered our homes, books and reading went out, if not immediately, then as generation after generation relied more and more on visual images and less and less on written (and perhaps, even spoken) words.

As I have said before, I was one of the first non-military persons in the Hawaii to experience computer conferencing (as it was then called) when I was invited to participate in an NSF-sponsored project, called the "Electronic Information Exchange System" (EIES), run by Murray Turoff of the New Jersey Institute for Technology. I also spent some time accessing the PLATO system at the University of Illinois, so when computer conferencing became possible via the University of Hawaii computer system, I was one of the first instructors to require students in my classes also to participate in an online chatroom. The first several times I did that, students would drop my class on the first day, expressing fear of computers and no desire to use them. But that soon faded, and so did my control of my classes. I was astonished when students would express opinions about my lectures and activities that students had never been able to express before. They were not flaming. They just were learning things I didn't know I was teaching while ignoring things I thought were important. Moreover, I quickly became just another guy trying to get a word in edgewise in the online portion of the class. It revolutionized not only my teaching, but also and mainly my image of myself as "teacher" and they as "learners". I had always prided myself in encouraging free and open class discussions, but what was said and unsaid face-to-face (especially in Hawaii) was totally different from what folks said online! I realized this meant the death of expertise and authority generally which social media have now made abundantly clear and irrevocable.

These experiences also reinforced my assumption that print literacy would decline and visual literacy increase. I imagined the future of communication at that time to be what in fact became YouTube and TikTok, and beyond to brain-to-brain and brain-to-AI direct communication.

I was astonishingly completely unaware that Evelyn Berezin had already invented and begun selling the first "word processor" (McFadden 2018). I first heard someone utter "word processor" when George Kent came to my office in fall 1976 just after I had returned from Canada and said I should come downstairs to get a glimpse of the future. Someone was demonstrating a word processor, he said. The phrase conveyed no meaning to me at all. How does one "process" words? I soon learned, as someone put a Wang 1200 WPS through its paces. The copy and paste/cut and paste functions blew me away.

But my heart sank because I realized the word was not out after all. Indeed, I curse An Wang and his gang and successors for postponing the death of print and the rise of visual literacy by fifty years. I thank Apple for doing what it could to keep visuality alive, though barely breathing, and thank also Sony Portapak (upon which I initially relied), YouTube, TikTok and electronic games for leading successive generations from the dungeons of literacy to the mountain tops of visuality. But there is still a long, dusty road ahead of us, as manifest by a Supreme Court that believes 21st America should be ruled by their interpretation of the meaning in 1787 of the handwritten words on a document fretfully pasted together during a long hot summer in Philadelphia. (The Founding Fathers used the hand-powered printing press each night to print and distribute each day's drafts and revisions, but knew the proper way to present the official version to the public was to write the Constitution by hand, not by an impersonal machine).

One of the topics at the Mediacy Fair I mentioned above was "Does reading rot the brain?" in which I and others argued that literacy does indeed rot the brain. You don't have dyslexia if you don't have disabling lexia to begin with. Or at least languages with alphabets, like English, that require linear processing of abstract signs to read and write seem to rot some brains, while languages based on holistic, meaningful, visual imagery, like Chinese, may not.

And then there was the controversy over the impact of electronic games.

Marshall McLuhan wondered what education would be like if television had been invented before the printing press. He said that sending a child to school set back her education by seven years; that she had learned more useful things about the world by watching TV at home than she could ever make up for in boring book-based schools.

Similarly, Steven Johnson imagined what education would be like if electronic games had been invented before the printing press, and it is not a pretty picture:

> Reading books chronically understimulates the senses. Unlike the longstanding tradition of gameplaying—which engages the child in a vivid, three-dimensional world filled with moving images and musical soundscapes, navigated and controlled with complex muscular movements—books are simply a barren string of words on the page. Only a small portion of the brain devoted to processing written language is activated during reading, while games engage the full range of the sensory and motor cortices.
>
> Books are also tragically isolating. While games have for many years engaged the young in complex social relationships with their peers, building and exploring words together, books force the child to sequester him or herself in a quiet space, shut off from interaction with other children. These new 'libraries' that have arisen in recent years to facilitate reading activities are a frightening sight: dozens of young children, normally so vivacious and socially interactive, sitting alone in cubicles, reading silently, oblivious to their peers.
>
> Many children enjoy reading books, of course, and no doubt some of the flights of fancy conveyed by reading have their escapist merits. But for a sizable percentage of the population, books are downright discriminatory. The reading craze of recent years cruelly taunts the 10 million Americans who suffer from dyslexia—a condition that didn't even exist as a condition until printed text came along to stigmatize its suffers.
>
> Perhaps the most dangerous property of these books is the fact that they follow a fixed linear path. You can't control their narratives in any fashion—you simply sit back and have the story dictated to you. For those of us raised on interactive narratives, this property may seem astonishing. Why would anyone want to embark on an adventure utterly choreographed by

> another person? But today's generation embarks on such adventures millions of times a day. This risks instilling a general passivity in our children, making them feel as though they're powerless to change their circumstances. Reading is not an active participatory process: it's a submissive one. The book readers of the younger generation are learning to 'follow the plot' instead of learning to lead (Johnson 2005: 19–20).

Or, to quote McLuhan one more time, "anyone who makes a distinction between education and entertainment doesn't know the first thing about either".

## References

Dator, Jim. 2019. Indivollectivity. *Critical Muslim* 29(January–March): 163–166.

Dator, Jim, John Sweeney, and Aubrey Yee. 2014. *Mutative media: Communication technologies and power relations in the past, present, and futures*. New York: Springer Press.

Johnson, Steven. 2005. *Everything bad is good for you: How today's popular culture is actually making us smarter*. New York: Riverhead Books.

Levine, Arthur. 2012. EdLife. *The New York Times*, November 4: 6.

McFadden, Robert D. 2018. Evelyn Berezin, computer pioneer who built first word processor, dies at 93. *New York Times*, December 10: A25.

Randolph, John. 1964. The senior partner. *Japan Quarterly*, January–March.

Worth, Sol, and John Adair. 1972. *Through Navajo eyes: An exploration in film communication and anthropology*. Albuquerque: University of New Mexico Press.

# Chapter 13
# *Ad Astra*! Sort of...

**Abstract** A lengthy discussion of the past, present and possible futures of humans, cyborgs, artilects, and posthumans on Earth and especially not-Earth. Adapting human becomings for life in every niche in the cosmos—one of the most profound shifts towards fluid forms and identities of all.

**Keywords** Adaptation · Afrofuturism · Analogs · Becomings · China · Culture · Cyborgs · Disability · Dreams · Dystopia · Entrepreneurs · Eutopia · Evolution · Gender · Governance · Humans in space · Identity · India · Intelligent environments · Military · Nation · *Ningen* · Not-Earth · Queer · Posthumans · Religion · Rockets · Science · Space fiction · Space history · Space junk · Spaceflight · Tenrikyo · Transhumans · USSR

I have been arguing throughout that each of us should carefully consider our identity; that we should create places and processes where people have the opportunity and expectation to contemplate what they want to become next, and enabled to move on if they wish to or remain as they are to the extent such stasis is possible in a profoundly weirding world.

That most certainly means we should all reconsider what it means to be "human" and whether or not it might be more interesting to expand that meaning to incorporate something else—something transhuman, cyborganic, or posthuman; indeed whether we wish to continue with the Enlightenment pretense that we are solitary individuals—人—or are in fact already interdependent social human becomings—人間. Perhaps we should strive to lose many features of western individualism and become participants in the hivemind? In truth, each of us and all of humanity from its beginning have never been the kind of rugged individuals many Americans, and especially libertarians, imagine. We are each ingredients in a pot of soup—tasty, healthful, pleasing together; tasteless, limited, pretty boring by ourselves. It is the soup that is primary, not each ingredient, but each ingredient contributes importantly to it. Why not go all the way: put the soup in a blender and turn on the switch, losing ourselves in the way Buddhism and other possibilities of being suggest, and that the Internet and beyond to the Noosphere makes plausible?

While it is clear that for many people the most important identity of all is being a human, for many others not only is the definition of "human" dynamic and malleable



J. Dator, *Beyond Identities: Human Becomings in Weirding Worlds*, Anticipation Science 7, https://doi.org/10.1007/978-3-031-11732-9_13

but it is also discardable—a useful shell for a while, but nothing to keep forever. Indeed, the nature of humans is to change forever—from pre-embryo to post-ashes—not to be stuck in the same form for eternity. So of course, a next step might be to merge with other humans, machines, life forms, the Earth, and the cosmos.

Getting out of the cradle, exploring the neighborhood while going cautiously and responsibly where no humans have ever gone is essential. But humans can never truly slip the surly bonds of Earth and ever go very far into the void in our current biological forms. We must evolve and design forms that are as suitable for the many niches of not-Earth as the human body was for Earth—when we emerged into a very different Earth long ago.

As the anthropologist Ben Finney said, "ever since our ancestors started using tools to survive and eventually flourish in new environments, the pattern of evolution by cultural as well as biological adaptation has been underway. Although the prospect of traveling and living in space might seem 'unnatural' to many, it would represent a logical extension to the technological path our ancestors have been following for some 5 million years." "If our descendants spread far and wide through space, the forces of evolution now braked on Earth will be released once more." "Human evolution in space will hardly be limited to the birth of one new species. Space is not a single environment…. There are innumerable environments out there providing countless niches to exploit, first by humans [and our accompanying artilects], and then by the multitudinous descendant species. By expanding through space we will be embarking on an adventure that will spread an explosive speciation of intelligent life as far as technology or limits placed by competing life forms originating elsewhere will allow" (Finney and Jones 1985: 335).

Finney is relying entirely on the natural forces of evolution to do their thing and transform *homosapiens, sapiens*, into successful spacefarers. He was always somewhat suspicious of claims that humans could or should be purposely modified to speed up the evolutionary process, and in any event was writing just as such claims were gaining plausibility and champions of their reality. But in addition to natural evolutionary forces now "braked on Earth" being unleashed as humans in their current form move into various immediate and eventually more distant niches in space, so also artilects, cyborgs and Pearsons *homo whateverus*—bionanomechanical "multitudinous descendant species"—may fill the cosmos.

Many years ago, Jerry Glenn and George Robinson issued a "Declaration of Independence for Spacekind as it Separates from Earthkind." Here is a portion of that Declaration. You may recognize its stylistic source:

> "When in the course of human evolution it becomes necessary
> for progeny to dissolve the political and biological bonds which have
> connected them with their progenitors, and to assume among the
> powers of the solar system and galaxy the separate and equal station to
> which the Laws of Nature and their Creator entitle them, a decent
> respect to the opinions of Earthkind requires that they should declare
> the causes which impel them to their separation into Spacekind.
> "We hold these truths to be self-evident, that Earthkind and

> Spacekind are created equal to their own respective environments, that once having been raised above their biological origins to a recognizable level of sentience and sapience they are endowed by their Creator with certain inalienable rights, and that among these rights are survival, freedom of thought and expression, and the evolution of individual and community knowledge."
>
> "We, therefore, the representatives of space migrants, space communities, and Spacekind descendants of Earthkind..., do, in the name and by the authority of Spacekind settled and living in space communities, solemnly publish and declare that their communities and their inhabitants are free and independent; that they are absolved from all allegiance to the governments and organizations of Earth; and that all political and ideological subservience of Spacekind to Earthkind is and ought to be totally dissolved.... And for the support of this Declaration, with a firm reliance on the protection offered through the Creative Intent, we mutually pledge to each other our lives, our fortunes, and our Sacred Honor" (Glenn and Robinson 1978: 202–207)

This is all so heroic and inspiring, reflecting the spirit of the early days of the erstwhile "space age". But why and how did the space age ever happen, and what might be its futures? (Dator 2012).

As we have seen, many cultures have stories about beings from not-Earth visiting Earth, as well as about people from Earth visiting the Moon, Mars, and elsewhere. For centuries, humans flying anywhere was only a dream. But a strong, compelling, widespread dream. What seemed to turn that dream into reality?

Leonardo de Vinci (1452–1519) of what is now Italy, designed flying ships but did not build or fly any. Some especially gifted—or daring—people actually tried to fly. It is said that Wan Hu tried, unsuccessfully, to fly with 47 rockets and two kites attached to a chair, during the Ming Dynasty in China (1368–1644). The Wright Brothers from the United States may have been the first humans to fly a machine-powered, heavier-than-air craft at Kitty Hawk, North Carolina, in 1903. The Soviet Union was the first country to succeed in actually placing a human in orbit in space—Yuri Gagarin in 1961—while the first man to land on the Moon was the American, Neil Armstrong in 1969.

Why did the millennial-long dreams of flying in space finally come true in the middle of the 20th Century?

First of all, from the late fifteenth century on, stories about a few intrepid western men voyaging across vast trackless oceans, invading previously unknown lands, exterminating many of the strange people on them while importing other westerners to claim the opened lands as their own (typically along with slaves from Africa and elsewhere to work them), gave some western men the experience and skills, along with a sense of god-given superiority and righteousness, that they could overcome any barrier and defeat any enemy. Later, the scientific-industrial revolution of the 19th Century produced so many new things so very rapidly that many people began to imagine that the future—their personal future and that of humanity generally—might be fundamentally different from, and in fact permanently better than, the present or any time in the past. This produced a new belief in perpetual and unstoppable "progress" and "development". This belief became the dogma—not to say the eventual insanity—of "continued economic growth:" the belief and experience of continual social and environmental change that led people to begin to write stories, compose music, paint pictures about the future as well as create and staff institutions that would keep the economy growing—financial systems to be sure but primarily educational, communication, governmental and religious institutions and practices.

America after World War II was vastly different from prewar America in terms of national pride, self-confidence and eagerness to undertake any task no matter how difficult, secure in the belief they would succeed. Science, technology, and a "can do" spirit encouraged Americans to engage in many ventures they probably would not have undertaken before.

The personal and social experiences of rapid social and environmental change led more and more people to expect novelties, to imagine things that had been impossible before, and then spurred some to try to create technologies that could make the dreams come true. World Fairs in the nineteenth and twentieth centuries played a big part in this as I showed with my own personal example from the New York World Fair of 1939–40. Also, as we saw, the world's most important and influential space fiction writer, the Frenchman, Jules Verne, and the British writer, H. G. Wells, and others, were specifically mentioned by almost all the early space pioneers as inspirations for their scientific and technological work. But the popular press was full of books and magazines replete with enticing depictions of utopias on Earth as well.

Later, pulp science fiction and space fiction movies, television shows, and popular music inspired generations of people all over the world to dream of space and to support space activities by their nations. But such stories could not have arisen and become as popular as they did without the technologies that created the experience of permanent social and environmental change, leading people everywhere to begin to wonder what new things might lie ahead, just over the horizon and beyond.

Fireworks and firework rockets were developed by the Chinese and their neighbors for fun, spectacular displays, and religious rituals often associated with the time of year crops were planted or reaped. Chinese developed festival fireworks around 600CE while rocket-powered weapons followed around 1000CE. This "dual use" of rockets and other technologies first for playful and then for military purposes (or vice versa) has been a continuing feature of space history. The first confirmed use of war rockets was in 1232CE when China used them against the Mongols. Mongols, in turn, used rockets against Poland in 1241 and against Baghdad in 1258. In 1249, Muslim armies used rocket-powered projectiles during the Seventh Crusade. By the end of the 13th Century, rocket weapons were known in Japan, Java, Korea, and India. Knowledge of rocketry spread quickly throughout Asia and into Europe at the same time.

Johannes Kepler (1571–1630), and Isaac Newton (1642–1727), so crucial to all aspects of modern science and technology, laid the scientific foundations for rocketry and rocket-propelled space travel. Kepler, a German mathematician, published three fundamental laws of planetary motion in 1609–1619. Isaac Newton, an English mathematician, established the basic laws of force, motion, and gravitation in his *Philosophiae Naturalis Principia Mathematica* (Mathematical Principles of Natural Philosophy), in 1687. Newton's Third Law of Motion was the first scientific definition of the reaction principle.

Later, Hyder Ali (1781) and Tipu Sultan (1792–1799) used advanced war rockets against the colonizing British in India. Indeed, it was the success of the Indian war rockets that inspired Colonel William Congreve to develop a new war rocket for the British in 1804. These rockets were used as signals and as artillery. There is a phrase in the *National Anthem* of the United States about the "rockets' red glare, the bombs bursting in air" that refers to England's use of Congreve's rockets against America during the War of 1812. European and US armies quickly adopted Congreve-style rockets and worked to improve them.

The Russian Konstantin Tsiolkovsky (1857–1934) might very well be considered the most important single individual in making space flight both a popular dream and a reality. A man of extraordinary vision and imagination as well as of deep knowledge and reflection, he provided a solid scientific basis for space travel. He not only wrote inspiring but plausible science fiction, but also made the first designs for multi-stage rockets and space stations; designed life support systems and space suits; and explained the feasibility of satellites and of solar energy. He mathematically established all the basic laws of space flight, demonstrating that liquid fuel rockets would have the thrust necessary to put a rocket into Earth orbit, or to journey to planets. In 1903 he published *The Exploration of Space with Reactive Devices*, the first major work on astronautics. It included the first presentation of the 'rocket equation'.

Tsiolkovsky was driven by the belief that space exploration was an absolute requirement for humanity that would lead to the colonisation of the Solar System. He wrote extensively in justification of that belief, arguing it was a moral imperative of humanity to move quickly into space.

Tsiolkovsky was as handicapped and marginal a person as one can imagine. He did not come from an influential family. He did not go to good schools and get good grades. He did not live in a population center where he could get and give inspiration to and from others. Rather, Tsiolkovsky was the son of a poor wood gatherer who entered Russia illegally from Poland—and illegal immigrant! His mother died when he was young and his father was seldom at home. He was deafened at an early age and never received any formal academic degrees of any kind from any school. He spent most of his life in provincial towns like Kaluga and Vyatka far from laboratories and good books. But he left Vyatka at 16 and went to Moscow in order to study in the libraries there.

His brilliance and desire to learn caught the attention of Nikolai Fyodorov, who was the chief cataloguer at the Chertkov Library, and himself one of Russia's greatest geniuses and eccentrics. Fyodorov enabled Tsiolkovsky to study everything that was being taught in the science and math departments of Moscow University. More importantly, Fyodorov taught Tsiolkovsky Fyodorov's own "philosophy of the common task"—the belief that humans should stop wasting time and intelligence killing one another, or buying and selling trinkets, and learn to transcend the chaos and chance of the cosmos—especially to transcend death. In doing that, humans should move off the Earth and go to other planets for new sources of food, energy, materials, and living room.

It was this philosophy—this almost religious dream—that inspired Tsiolkovsky to do all the extraordinary pioneering work he did in developing the basis for all scientific, technological, and social aspects of human space flight before anyone else did, and well before the technology itself enabled it. In a very important sense, only a marginal but intelligent person such as himself could have dreamed the impossible dreams he did, and then had the patience and ability to figure out how the dreams could eventually come true. He did not have parents, teachers, priests, companions, and an entire culture telling him his dreams were stupid—or evil—as most of us do. Neither was he told he should become a dentist or a lawyer or an insurance salesman, and stop lying about the house reading and reckoning. He could dream freely and plan brilliantly without crippling ridicule or stifling support.

There were a few other people like Tsiolkovsky in other countries who often initially were also lonely eccentrics working in ignorance of each other who laid the basis for eventual spaceflight. A small international space flight movement began to form in the 1920s and 30s. Enthusiasts in many countries formed space travel and rocket societies, carrying out theoretical and practical research in rocketry and spaceflight. Some of these groups made very significant contributions to the development of rocketry and astronautics, and nurtured the talents of future spaceflight engineers. But they all were ridiculed as silly hobbyists and dreamers by most other people at the time and received little attention or financial support.

One of the most important early spaceflight societies was *Verein für Raumschiffart* (VfR: Society for Space Ship Travel) formed in 1927. Many members went on later to develop rocket technology during and after World War Two. Most prominent was Wernher von Braun.

In 1921, the Gas Dynamics Laboratory (GDL) for rocket research was established. Ten years later, the Group for Study of Reactive Devices (GIRD) was formed. It designed and launched the USSR's first liquid-fuel rocket. Among GIRD's members was Sergei Korolev who later became the leading figure in the Soviet space program.

The Treaty of Versailles, after the First World War, severely limited the kinds and amount of weapons Germany could have. But as nationalism and hypernationalism began to rise, rapidly inspired by and inspiring Hitler, rockets were seen as a way around the limitations of the Treaty. Werhner von Braun and other VfR members were employed by the German army to work on rocket projects. In 1937, the German Army created a rocket research center on the Baltic Sea island of Peenemünde. Largely using Jewish and other slave labor, they developed the A-4 (also called V-2: Vergeltungswaffe 2 or vengeance weapon). It was a tremendous leap forward in rocket technology. The V-2 established the basic design concepts for rocket motors, fuel systems, guidance, and steering that remain at the heart of today's rockets. More than 3,000 V-2 s were launched by the Germans against their enemies during the last stages of the war, causing 2,700 casualties in Britain alone. Though it was a stunning technological breakthrough, the V-2 was not a successful weapon per se: it could not be guided with accuracy, and its complex technology was unreliable. Its main military value was as a terror weapon. Unlike bombs from airplanes that could be seen and heard well in advance so that evacuation to shelters was possible, V-2 were unnoticeable until seconds before they exploded. The screaming brief sound they made before impact was horrifying.

The V-2 was also enormously important as a harbinger of space flight and, more ominously, of the possibilities of intercontinental war. The V-2 was the world's first long-range missile. After the War, it became the prototype for the first ICBMs that played key roles during the Cold War as potential deliverers of world-destroying nuclear weapons across the planet in the twinkling of an eye. V-2 derived missiles were also developed into the first space launch vehicles.

It is clear that rockets and space flight were minor concerns for most governments and businesses until it was seen they could be used as mass killing machines. Suddenly millions of dollars were poured into their development for military purposes. Only the Germans were able to bring military rockets online during World War II, but did so too late for their success: the war was basically already decided in Europe before the V-2 s could do much good—or bad.

It is worth noting that during the Second World War, even though efforts were made in the United States to develop rockets, the real money and talent there was spent on developing the atomic bomb. The first usable atomic bombs were also developed in the latter stages of the war, and were used by the United States against Japan—devastating the cities and decimating the inhabitants of Hiroshima and Nagasaki. It is further worth noting that many of the people who were instrumental in developing the science and technology necessary to create those bombs for the United States were from Germany—for the most part Jews who had fled from Germany and Europe to the US to escape the anti-Jewish wrath of Hitler and the Nazis. In contrast, the developers of the German military rockets, like Wehrner von Braun, were "Aryans"—not Jews.

The German rockets carried conventional explosives. The Germans had no atomic bombs. On the other hand, the Americans used conventional airplanes to deliver atomic bombs to Japan. America had no long-range rockets anywhere near the stage of development that the Germans did. Imagine what the world would be like now if Germany—or America—or both had had both rockets and atomic bombs in time to use them in World War II. Dreams, even bad dreams of hatred and prejudice, drive human activities, and we living now can only consider ourselves fortunate that Nazi ethnic prejudice against Jews contributed to these two mighty tools—nuclear bombs and long-range rockets—not being joined in Nazi or American hands before the war was over.

Nonetheless, after World War Two, nations rushed to develop their own long-range missiles. Military advantage dominated rocket development—not the desire for space flight. As the Cold War developed during the 1940s and 50s, creating ICBMs to deliver nuclear weapons to any spot on the globe became part of the strategic planning of all major nations. That received top priority in both the US and the USSR, with massive resources being poured into missile development and advanced nuclear bombs.

Out of the libraries come the killers.
Mothers stand despondently waiting,
Hugging their children and searching the sky,
Looking for the latest inventions of professors.
Engineers sit hunched over their drawings:
One figure wrong, and the enemy's cities remain undestroyed.

Attributed to Bertoldt Brecht (http://www.rosenfels.org/ToEsther)

Indeed, as World Was Two was ending, the Soviet's, moving to Germany from the east, and Americans and its allies moving into Germany from the south and west, rushed to capture German rocket experts and technology. As it turned out, the Americans and their allies got to the best ones first, and quickly removed most of the scientists, engineers, and rocket parts and technologies to the United States. There was not much left for the Soviet's to take back to the USSR as both countries began investing heavily in rocket technologies for war.

Sergei Korolev (1906–1966) who had been a member of GIRD became the famous (and for many years anonymous) "Chief Designer" of the Soviet space program. From the end of WW2, Korolev oversaw development of the USSR's long-range missile program. He was the driving force behind the USSR space program, quietly appropriating military technologies for space flight, and finally gaining official support for developing space technologies. His efforts were enormous, carried out almost in secret, not only from the outside world, but also within Russia, until his rockets were finally and successful launched. It is indeed uncertain how the competition for space dominance between the US and the USSR would have turned out had he not died suddenly, during routine surgery, in 1966, after which the Soviet program floundered.

Wernher von Braun (1912–1977) was the leading figure in the development of the V-2 for Germany. After the war he was brought to the US along with many of his German co-workers to lead the development of America's missile programs. He did so spectacularly, technologically, enabling the US space program to develop very rapidly based on the advanced knowledge and experience he brought with him from Germany. Von Braun was also an extremely effective—indeed, charismatic—populariser of space travel in the US, his thick German accent and tall, noble bearing fascinating and attracting more Americans than were repulsed by his Nazi associations.

Scientific upper atmospheric research provided the trigger for the space race. In the mid-1950s, the world scientific community proposed the International Geophysical Year (IGY) to investigate the Earth's relationship with the space environment. Both the USSR and the United States declared that they would build scientific satellites as part of their research program.

October 4, 1957, unexpectedly and without previous fanfare, the Soviet Union launched the world's first artificial satellite, Sputnik-1. The space age had begun. More than 50 times larger than America's proposed satellite, Vanguard, Sputnik-1 was a stupendous technological and propaganda victory for the Soviet Union. Startled world reaction to Sputnik was spontaneous and frenzied. The Russians were suddenly leaping ahead. On November 3, Sputnik-2 placed the first animal into orbit, the lovable dog, Laika. She proved that it was possible for living things to survive the rigors of launch and the conditions of weightlessness. The Soviets had no plans to return her to Earth and her heroic death in space was mourned by millions worldwide.

After the tremendous propaganda victory of Sputnik-1, the US government authorized a crash program to develop the Explorer satellite as alternative to the troubled Vanguard project. Von Braun and his associates, including scientists from the Jet Propulsion Laboratory in California, constructed and launched Explorer 1, on January 31, 1958, from Cape Canaveral.

The United States had expected that it would launch the world's first satellite. When the Soviet Union achieved this distinction instead, it sparked an extreme panic reaction in the Cold War environment of the time. If the USSR could put a satellite into orbit, it was assumed they also had the technology to attack the US from space. This fear rapidly led to a 'space race' between the two superpowers. Americans responded with frantic determination. American President, John F. Kennedy, famously declared in May 1961 that before the decade was out, America would land a man on the Moon—and bring him back to Earth. A bold statement! This was something that was technologically impossible when he announced it (especially the "bringing back" part). America had to make good on his boast, or suffer even greater loses of respect and leadership.

The space race was on with competitive national prestige and security now a driver that led to sputniks and humans going into space. Political propaganda quickly associated space achievements with ideological superiority: "my political system is better than your political system because my space technology is better"! The US and USSR vied with each other in a race to achieve status-conferring space 'firsts'. Initially the Russians beat America in almost every aspect in the race to the Moon. Working under a cloak of heavy secrecy and with substantial funding and human resources, Russia achieved first after first, while America's failures and belated successes were widely-known and ridiculed.

Some Soviet firsts in the space race include:

1957-First artificial satellite (Sputnik 1)

1957-First living creature in orbit (Laika the dog, Sputnik 2)

1958-First satellite to weigh more than one ton (Sputnik 3)

1959-First successful lunar probes (Luna 1, 2, 3)

1961-First person in space (Yuri Gagarin, Vostok 1)

1961-First 24 h space mission (Gherman Titov, Vostok 2)

1962 First 'twin' flight with two manned spacecraft in orbit at same time (Vostok 3 and 4)

1963 First woman in space (Valentina Tereshkova, Vostok 6)

1964 First multiple-crew spaceflight (Voskhod 1 with a crew of three)

1965 First "space walk" (Alexei Leonov, Voskhod 2)

and many, many more.

It seemed that Russia was vastly superior to America in the science, technology, and human abilities necessary for space flight. Nonetheless, the Russians failed to attain what many believed to be the main prize in the race. The Americans were the first to land men on the Moon in July 1969, and to return them safely back to Earth, as Kennedy had promised. Americans were ecstatic. Everything was space age this and space age that. "The Future" and "Space" became synonymous for many people, and indeed America continued to send men to and from the Moon until 1972 when political will and public enthusiasm failed. After Americans reached the Moon, the Soviets gave up the race and turned their attention to other goals, all scientifically and perhaps strategically important, but none as romantic or appealing as the race to the Moon.

To achieve the Moon landing, incremental steps to a permanent human presence in space (set out as far back as Tsiolkovsky) were pushed aside in a headlong rush for geopolitical dominance and propaganda glory. The race to the Moon did not generate the space technology and infrastructure necessary to support a permanent human presence in space, such as space stations and cheap access to space with versatile crew and cargo vehicles. Thus, after the Apollo landings, the pace of space technology development slowed markedly and reverted to a more 'normal' type of technological change. Space activities took a step backward, arguably providing time and resources needed to develop step by step the necessary space infrastructure needed for a permanent human presence in space. But even this did not happen in a systemic, logical way.

The space race was over.

In 1979, Arthur C. Clarke referred to spaceflight as a "technological mutation that should not have occurred until the 21st Century". He meant that the technologies (hardware, software and orgware) essential for the missions politically set had to be developed much sooner than would normally have been the case. They were not optimal for their task. Ordinary technological advances are the products of a gradual process of incremental innovation and improvement. Revolutionary technological change may be brought about by social and political decisions operating outside conventional market and scientific processes, but then prove to be unsustainable. Spaceflight was a revolutionary technological change that arrived, driven by geopolitical factors, well before the world was ready to exploit its full potential and before the technology achieved the maturity needed for the job.

Many people tell the story of the space race as though it was only between the US and the USSR. Though less well-known abroad, many other countries were part of the space race too, and on their own; not necessarily as allies of one cold warring nation or the other. The "father of Chinese rocketry" was Qian Xuesen. Qian earned a PhD from the California Institute of Technology and helped the US develop jet propulsion and the atomic bomb during World War II. However, during one of the many "witch hunts" that plague the United States then and now, Qian was expelled from the US in 1955, and returned to China, where he become director of China's rocket research. Fully supported by Chairman Mao, China opened its first missile and rocket research facility in 1956. This is another of the ironies of the space age: political aims and ideological beliefs both drove and handicapped national aims for the development of space and domination of the Earth. Initially using Soviet technology, China launched its own R-1 rocket in November 1960, while placing its first satellite in Earth orbit on its own "Long March" rocket on April 24, 1970, broadcasting the song, "The East is Red". For over 200 hundred years, "red" meant left (liberal, progressive, socialist, communist) and "blue" meant right (royal—i.e., blue bloods—conservative, corporate, fascist), derived from the seating of political parties in the French National Assembly in 1789 on the left and the right sides of the chamber, so by proclaiming "The East is Red," China was declaring victory for Communism with its successful rocket launch. How and why the meaning of the colors red and blue got switched in the US, apparently by television stations broadcasting election returns, is beyond me. In any event, China has continued to make enormous advances in space activities in all areas, rivaling and perhaps soon exceeding the United States.

In 1962, with the support of Prime Minister Nehru, Vikram Sarabahi became the father of the Indian space program. With technical support from both the US and the USSR, Sarabahi was less interested in weapons and races to the Moon. He saw Earth-orbiting satellites as tools for national development that provided communications, remote sensing, weather forecasting, and both educational and popular television broadcasting in support of national goals. The first Indian satellites were launched on Russian rockets in 1975 and 1979. India also has steadily and impressively developed its national space program.

The French space program, Centre National d' Etudes Spatiales (CNES), was established under President Charles de Gaulle in 1961. The Diamant rocket, launched in 1965, was the first satellite launcher not built by either the US or the USSR. Other nations developed space program during this same period as well.

## 13.1 Identities in Space

All of the first people in space were either from the Soviet Union or the United States, and some of their political allies. China, most importantly, became the third country to put a man in space with its own rocket in October 2003. Yang Liwei in September 2011 became the first Chinese on the Moon. India may launch its first crewed spaceflight in 2023. And there are the odd rich space tourists now. As of 2021, people from 41 countries have gone into space, almost all from the US and Russia distantly followed by Japan, Germany, China, France, Canada, Italy and others.

While all nations have some kind of space programs, most are focussed on satelitte applications. Very few are concerned with human space flight. Moreover, only eleven nations are "spacefaring", meaning they have the ability to send objects into space using their own launchers. Many more are racing towards spacefaring capacity, in the meantime renting space from spacefaring nations and companies in order to launch their satellites into orbit. Indeed the big business in space is launching satellites and providing programming that can then, for a price, be beamed from one‘s satellite back to specified spots on Earth.

The United States, the USSR, and China established their space programs long ago for military and/or nationalistic purposes, though they also typically had some kind of formal separation between the far more heavily-funded amd secret military side, and the more celebrated civilian side. Military activities were often hidden in secrecy with the extent of their surveillance activities and the weaponization of space heavily obscured. However in 2019 the US created the US Space Force, a specifically menacing name. Space Guard or Space Corps might have been much better. Other nations had little choice but to respond in kind with their own space military unmasked. In some ways, "space force" perhaps turned out to have been a good choice because it quickly morphed into "space farce" in comedy and parody. I'll take the parody over the very real possibility of future space wars which, aside from all the other devastations, might so greatly increase orbiting space junk that further launches from Earth will be untenable unless vehicles impervious to collisions with the debris are developed.

While there was never anything impelling space programs philosophically or spiritually equivalent to that of Russia's Fyodorov or Tsiolkovsky‘s "Philosophy of the Common Task", religious symbols and memes have played a large part in America‘s space program from the beginning. Perhaps the first and most famous was when Frank Borman read from the book of *Genesis* on Christmas Eve 1968 as *Apollo 8* circled the Moon. Buzz Aldrin performed the Presbyterian service of Holy Communion on *Apollo 11,* and Jeff Hoffman (the first Jew in space) on SST61 (1993) celebrated the Jewish Hanukah with menorah and dreidel, but, thankfully, no burning candles.

Deana Weibel states that "The United States is a country, then, that sees itself, at least in terms of its historical mythology, as following the dictates of God and being rewarded in these pursuits." She examines a video produced by the new Space Force and observes,

> While the language in this section is not quite as iconic as famous military slogans like "Be all that you can be" or "We do more before 9 am than most people do all day," it is still practical, realistic, and militaristic in character. The narrator intones: "Some people look to the stars and ask, 'What if?' Our job is to have an answer. We have to imagine what will be imagined, plan for what's possible while it's still impossible." The message is one of optimism and determination. The second part of the video continues with innovative military imagery, while the narrator's language becomes philosophical and perhaps even metaphysical. "Maybe," the voice actor suggests, "you weren't put here just to ask the questions. Maybe you were put here to be the answer. Maybe your purpose on this planet isn't on this planet." The idea that the listener was "put" here, of course, implies that someone or something, some kind of higher power, decided to put the listener here. The implication is that the listener, if he or she joins the Space Force, will have the potential to play a messianic role, one of salvation, although the video doesn't clarify whether it is the United States or the Earth that needs saving. (Weibel 2020)

Former Vice President Mike Pence frequently used religious language, including in his support of the space force. For example, "As President Trump has said, in his words, 'It is America's destiny to be the leader amongst nations on our adventure into the great unknown'. And today we begin the latest chapter of that adventure. But as we embark, let us have faith. Faith that, as the Old Book teaches us, that if we rise to the heavens, He will be there" (Weibel 2020).

Among the most overlooked constituencies for nongovernmental spaceflights are indeed various religious groups. Many of them possess great wealth and a vast membership that embraces all of the scientific, technical, social, political, and motivational talent necessary to sponsor significant space programs. If the evangelical Christian community were to grasp the opportunity to expand its witness beyond Earth it might very well do so quite successfully, and perhaps even in a way that could provoke a space race with other religious communities.

Tenrikyo is a Japanese religion that combines certain Buddhist, Shinto, and other features with perspectives of its own. According to a statement on the Tenrikyo website:

> The teachings of Tenrikyo are said to be direct revelations of God the Parent, who desires to save all human beings and thus enable them to live the Joyous Life.
>
> God not only created human beings and the world but has always been and will ever be the source of life and the sustainer of all things. The followers refer to God as God the Parent in the sense that God is the original Parent of human beings and, in prayer, glorify God as Tenri-O-no-Mikoto. According to Tenrikyo, "The universe is the body of God." It is thus said that the world is filled with God the Parent's workings.
>
> (Tenrikyo nd)

I am not certain why or when Tenrikyo first became interested in space exploration, but I do know that Ben Finney and I attended several meetings that Tenrikyo held in Japan on space-oriented themes. For example, in December 1986, the American Astronaut Russell Schweickart and myself gave public lectures as part of a "Pre-session" titled, "Toward the Cosmic Age" (Yamada 1988).

In July 1989, Schweickart, Russian Cosmonauts Oleg Makarov and Georgi Ivanov, and others participated at a seminar in Osaka on "Earth, Space and Human Beings." In June 1990, Ben Finney and Astronaut Charles Conrad were among the participants in a symposium in Tenri City on "Inner Cosmos and Outer Space: Searching for a New Cosmology."

The leading person behind these activities was Akio Inoue who was the Director of the Tenri Yamato Culture Congress. Among many other space-related activities, Mr. Inoue translated into Japanese parts of the diary kept by Cosmonaut Valentin V. Lebedev during his 201-day flight aboard Salyut 7 in 1982. Inoue was especially interested in Lebedev's accounts of his dreams in space, and in the forward to the excerpts, Mr. Inoue urged readers to look beyond the physical and technological aspects of space flight and to focus more on its spiritual and religious dimensions and implications.

Similarly, in an English-language prospectus for a "Japan International Space Culture Congress," Inoue wrote:

> In further pursuing space development, we need now to consider ever more intensely such basic questions as the relation between man and space, between the universe and the evolution of life forms, as well as why man exists in the universe at all. Such profound questions must be asked along with those concerned with the material sides of scientific and technological development....In Japan, for example, space research tends to over-emphasize the technological aspects, and to neglect such important social and psychological questions as the changes in human consciousness and society that space development is sure to bring about. These questions need to be as eagerly pursued as are those solely concerned with the technology of space exploration.

I am not sure Tenrikyo has continued or expanded these space activities, but they are illustrative of what we might expect from religious groups as spaceflight moves from the monopoly of governments into various private sectors.

For 400 years, the Roman Catholic Church's observartories have been part of the global community of astronomical observatories, contributing actively to the advancement of astronomy and related fields. I am not aware of any official plans underway for the Church to engage in actual space programs, but I do know that various clergy and laymen have spoken out concerning spaceflight and humanity's role in space—some strongly favoring, other opposing such ventures. The same is true for other demoninations of Christians.

Similarly, various Jews, Muslims and Hindu leaders have stated positions one way or another about spaceflight. In those countries where religion and governance are more tightly entwined than in the US or Europe, religious considerations are already part of official programs.

Religion is at least as great a motivator for human behavior as sex, so space religious tours might well rival space sex tours in popularity.

Modernity—the world that emerged in the late 17th Century and flowered in the 19th Century and early 20th Centuries—was created on the basis of scientific ideas as actualized by technologies derived from (and/or leading to) scientific principles. Although religion remained a major part of the beliefs and actions of most people, for a while it seemed that religious ideas were relics of older belief systems that were being replaced by ever-progressing new scientific discoveries and technological breakthroughs. Modern institutions of governance and education especially were to be based on the best modern science available. The very purpose of public schools and publicly-funded universities in the 19th Century in western Europe, north America, and Japan was to transform nations from agrarian, semi-feudal entities into urbanized, globally-influential powers producing the latest industrialized people, goods, and weapons.

The belief in the superiority and necessity of ever-improving science and technology reached its zenith with government-directed R&D projects in Germany, Japan, the Soviet Union, the United Kingdom, and the United States during and after the Second World War, exemplified by the atom bomb-producing Manhattan project on the one hand and the V-2 rockets and the subsequent Space Race on the other. Vannevar Bush's report to the US President in 1945—*Science, the Endless Frontier*—was the Bible of those beliefs and policies in the US. In the Soviet Union the term "Scientific-Technological Revolution" became a frequently-repeated mantra while from 1966, Radovan Richta's sixty-author volumes, *Civilizace na rozcestí—společenské a lidské souvislosti vědecko-technické revoluce* ("Civilization at the Crossroads: The social and human context of the scientific-technical revolution") broadened and deepened the context of science for society on both sides of the Iron Curtain.

As a consequence, after the war, governments of all advanced nations began for the first time pouring very substantial amounts of money into "R & D" (research and development) aimed at increasing scientific and technological knowledge and applications. Professors at certain universities and research institutes had the specific obligation of increasing scientific knowledge and tools in their field and, where possible, of finding practical applications for them. In order for a nation, a corporation, an individual to be successful, it was necessary to acquire and use the very latest scientific knowledge and tools. It was imperative constantly to produce new knowledge and new tools in order to get ahead, stay ahead, grow, and progress. There was no doubt about it.

Science in the Atomic Age/Space Age ruled during the 1950s, 60s, and70s. Communism, in the Soviet Union, Eastern Europe, and East Asia, was avowedly atheistic and scientistic. The first man in space, Yuri Gagarin, officially said he saw no God in space (though he allegedly told his adoring *babushkas* off the record that in fact he had). In contrast, as we saw earlier, during prime time television on Christmas Eve, 1968, the crew of Apollo 8 took turns reading the opening passages of the book of *Genesis* as they circled the Moon. NASA officially disapproved, and many Americans were exasperated by this archaic act. In some ways, the nadir of religion in the United States was reached with the cover of *Time* Magazine of April 8, 1966 that asked, in huge red letters, "Is God Dead?" with the clear implication that indeed he was.

However, if God was dead, that did not mean that most people had replaced religion with established scientific principles. For many years, the National Science Foundation has conducted extensive surveys of the scientific and technological knowledge of ordinary Americans. Without exception, most Americans demonstrate that they know and believe almost none of the fundamental principles of modern science; do not understand the scientific method or how to read statistics; and adhere to beliefs about humans and nature that science clearly rejects. Moreover, these surveys do not show that scientific and technological knowledge is increasing over time among Americans. To the contrary, nonscientific beliefs are increasing and faith in science declining.

Surveys in other parts of the world show similar patterns. However, it is very important to understand that, at the same time, surveys of religious beliefs also show that many devout people stoutly proclaim beliefs and exhibit behaviors that are contrary to the official teachings of their religion. Similarly, if not told the source, many Americans loudly denounce statements that derive from the US Constitution—especially the first ten amendments—as being the seditious ideas of communists. Laymen's/citizens' beliefs horrify theologians and political scientists, as well as scientists.

Getting folks to accept evolution has always been an uphill fight, and creation scientists seem to be winning the battle for recognition in many parts of the world. More recently, physicists' Big Bang theory of the creation of the universe has been attacked. Similarly, long before Covid-19 antivaxxers existed, the World Health Organizations said that the influence of various spiritual beliefs about health was in part responsible for the re-emergence globally of childhood diseases once thought eradicated by vaccinations or treated water. Highly likely these will spread if and when the pandemic ends.

Opposition to science and technology has not come only from religion. Indeed, there has been severe criticism of science and technology from the academic and cultural left. Studies stressing "the social construction of science" emerged in Europe and North America in the 1970s arguing that the claims of science are based not in "truth" but in ideology, politics, and claims of privilege and expertise. Science is said to be just one way of thinking among many others. It has no special claims on truth. The words of a scientist, even speaking as a scientist, are no more legitimate than the words of any other person who has opinions about a matter. Positivistic, reductionist science especially has been brought into serious question, and in some quarters, disrepute and rejection. Scientists are like anyone else. They lie, cheat, fudge the data, have special interests they want to protect and expand. There is a class system with a laboring class of technicians and graduate students at the bottom who do most of the drudge work while a small number of self-referential scientists take all the fame and money. Many horrible things have been done and are still done every day in the name of science. Ordinary folk must always challenge what a scientist says and not take his word on anything, people holding this position maintain.

More recently this claim is especially made around the world by, and on behalf of, some indigenous people's ways of knowing that science seeks to discredit and replace. Some argue that many indigenous beliefs are more wholistic, integrative, and indeed true than is reductionist science that only knows how to understand something by taking it apart and thus destroying its integral operation.

These leftist/humanist claims about the social construction of science were appropriated by energy and climate change deniers who say that the science in back of them is nothing but propaganda and the special pleading of scientists who want more grants for their so-called research.

In addition to science, as we have seen, many people have a love-hate relation with technology. While for some people, technology is good, or at least neutral, for others it is demonic: with every new technology, we move farther and farther away from our pristine Garden of Eden where the living was easy under God's providence. Technology has increasingly destroyed our environment, culture, and families. Rather than being a good thing to be encouraged, funding for more science and technology should stop. We don't need more technologically-caused social and environmental damage, we need less, more people are saying. Let's restore old family, community, cultural, and religious values along with the technologies and ways of knowing that supported them. Since a major reason for supporting basic scientific research is not the abstract love of knowledge but to lead to practical technological breakthroughs, the days of both science and technology may be numbered.

Most, if not all, of the founders of contemporary science were Christians who saw no inherent conflict in religious and scientific reasoning. However, many contemporary western scientists now see religion and science as operating in different spheres and appropriate for answering different questions. Many others think science should be an entirely secular activity completely divorced from religious faith.

However, centuries before there were scientists like Copernicus in Europe, there were Muslim scientists in southern Europe, Africa and the middle east. Contemporary science owes much to them. The position of most Muslim scientists then and now is similar to early western scientists: it is their Islamic faith that leads them to their scientific methods and discoveries. There is no true science without true faith. While entirely secular Muslim scientists exist, most insist their science derives from and is consistent with their faith. Many of these also maintain that "Islamic science" is distinct from and superior to "western science" because of the unity of faith and reason. Indeed Islamic science reveals scientific truths that western science has yet to accept, many would say. At the same time, as with Christianity, Judaism, Hinduism and others, so also there are Islamic fundamentalists who reject and condemn all scientific beliefs except those revealed by the Prophet.

Additiional steps towards governmental support of religion in the US occurred when the White House Office of Faith-Based and Community Initiatives was established by George W. Bush by Executive Order on January 29, 2001. The initiative enables various religious groups to obtain federal funds to provide certain social services offered by the federal government in what once would have been considered a violation of the separation of church and state clause of the First Amendment to the US Constitution. Much to the amazement of many of his supporters, President Barack Obama expanded the scope and funding. President Biden re-established the Office as the White House Office of Faith-Based *and Neighborhood Partnerships* (emphasis added), while a Center for Faith-Based and Neighborhood Partnerships in the Office of the Secretary of the Department of Housing and Urban Developed was created "to ensure that Faith-Based and Community Organizations, which form the bedrock of our society, have strong advocates within HUD, the White House and Partnership Offices throughout the Federal Government" (Center for Faith-Based and Neighborhood Partnerships 2022). So far, court cases have upheld the Initiative. Funding is also being offered to religious groups engaged in more than conventional social service work. Indeed, the US Supreme Court seems to be tending towards declaring in general that forms and venues of religious expression and practice long declared unconstitutiional are in fact permissable. Perhaps, eventually mandatory.

In contrast, some parts of the world—China and Japan most clearly—are completely enamored by conventional science and technology, and are doing what they can to increase the scholarly quality and quantity of their scientists.

At the same time, funding by the US and some other governments for science research and development continues its basically downward trajectory from the high of the 1960s, with some exceptions, generally sliding lower both in terms of absolute dollars and as a percentage of GDP. If science is understood to be just one belief system competing among many, 100% of all research funding may be given to faith-based organizations, of which science is just one. Religious fervor might replace scientific rigor as a main driver of space flight.

## 13.2 Small Sample

Only a few humans have ever been in space and only very, very recently in the context of human history. As of November 2021, 600 humans have flown in space since the first person, Yuri Gagarin, orbited the Earth in April 21, 1961. Of these 600 only 65 were women with Valentina Tereshkova of the Soviet Union being the first in 1963, though the second, Svetlana Savitskaya, did not fly until 1982, while the first American woman in space was Sally Ride in 1983. Since then the number of women in space has increased, but they are still very much a minority. To be sure, women played vital roles in the many research and administrative duties needed to get a few humans into space but even here they often were marginalized and harassed internally, and largely unknown and unsung within the space community much less the wider world.

All of the first astronauts, cosmonauts, and taikonauts were males. In the US, they were White, Anglo-Saxon, militarized test pilots, modestly-educated (BA/BS degrees), cisgendered, able-bodied and often farm boys, with the right stuff.

Eighty-eight of all astronauts have been White. Asians account for 7% of all astronauts, and African-Americans and Hispanics account for 2% each. In 1980, the Soviet Union became the first to send a Black man to space: Arnaldo Tamayo Méndez, a Cuban cosmonaut. It would take another three years until America sent its own Black man to space: Guion Bluford. Charles Bolden, a former astronaut, graduate from the US Naval Academy, and a retired major general in the U.S. Marine Corps served as NASA administrator from 2009 to 2017. In an interview he said that he was "'on a constant roller coaster of being very angry and then being hopeful,' not only about his experiences as an African-American man who has helped lead America in space, but also about how he feels about racism and violence against people of color today. 'We don't have enough representation in the astronaut service by women and minorities.' 'We have to have more representation … We've not had an African-American crewmember on the International Space Station, and that is long, long, long, long, long, long, long overdue.'" "The space station has been in orbit around Earth for 20 years "(Gohd 2020).

Episodes of serious concern and actions towards greater diversity within NASA come and go as they do in the American population generally and with the the predilections of presidents, other executive officers, and members of congress, as well as the heads of NASA itself. Sometimes there seems hope for sustained affirmative actions, but more often crude and blatant discrimination prevails. It has been said that the racial and gender discrimination in NASA is worse than any other branch of the US Federal government. Given the militarized history of US and other space programs, this is no surprise. Even though the US military, importantly, was the first major agency to end formal racial discrimination after World War II, it continues to resist as best it can the full inclusion of women, queers, trans and disabled persons into military service. In addition, many of NASA's installations are located in the US South where prejudice against equal opportunities for women, racial and gender minorities, and the disabled is entrenched and strong.

Given the cultures of America and most of the world in the early days of space flight, it is no surprise that as far as is currently known no early astronaut or cosmonaut was gay. It later turned out that the widely-revered first American woman in space, Sally Ride, was gay. This was not disclosed until her gay partner of twenty-seven years included that information in Ride's obituary in 2012. The second gay astronaut, Wendy Lawrence, also concealed her gay marriage until 2018, after she had retired from NASA following many years of distinguished service.

Anne McClain, who had been in an unpublicized gay marriage since 2014, was the first gay astronaut to have been outed while aboard the International Space Station, in 2018. McClain had been embroiled in squalid divorce proceedings and accused by her ex-wife of what might have been the first "space crime." She was cleared of all accusations.

No male astronaut on active duty or in retirement has come out as gay so far. There were rumors of gay astronauts in other countries, but none has been confirmed yet. Since 2016, an LGBTQ Special Emphasis Group within NASA has also been working beyond what federal law requires to create a more positive, equitable, and productive NASA work environment (French 2020).

Cameron Bess, who accompanied his billionaire father on Blue Origin's flight on December 11, 2021, may well be the first gay male in space and it was no secret. Indeed he intended his venture to be pacesetting and inspirational:

> Blue Origin @blueorigin · Dec 11, 2021
>
> Replying to @blueorigin
>
> Cameron Bess is a content creator with a passion for creating and expressing themselves in ways that can brighten a person's day. Cameron identifies as pansexual and is proud to represent marginalized communities and hopes their journey can inspire others. (Holmes 2021)

In addition to the LGBTQ Special Emphasis Group in NASA,

> The Out Astronaut Project addresses the under-representation of Lesbian, Gay, Bisexual, Transgender, and Queer (LGBTQ+) people in science and space. Sponsored by the International Institute for Astronautical Sciences, we highlight the contributions of LGBTQ+ members currently working in science and space and provide grants to promising LGBTQ+ students currently pursuing professions in space-related fields.
>
> "We believe that communities are empowered when they are represented. Astronauts inspire our youth, represent limitless possibilities, and serve as ambassadors to STEM. Our goal is to train and fly a member of the LGBTQ community as a scientist-astronaut. This person would be an inspiration to the LGBTQ community while enabling cutting-edge research." [https://outastronaut.org/]

But perhaps is not an issue of infusing space and the space community with queers. Perhaps space is queer already and it is Earth that it is odd man out, struggling to escape the bonds of oppressive normality, and breathe free.

"[Q]ueer online advice columnist, John Paul Brammer, answered the question, 'What is the unspoken bond that LGBTQ people have with space?' with 'we queers tend to know a lot about the ordeal of being perceived and then being negotiated into rigid taxonomies that weren't built with us in mind.'" "Emily Hunt, a queer trans woman studying star clusters for her PhD, [said] that she first became interested in the stars when she was around 7 years-old, after her parents showed her Venus transiting across the sun. She says she is humbled by space, and believes the universe is gay because it's inadvertently 'counterculture.'"

"Moiya McTier, a Black astrophysicist who identifies as bisexual and pansexual,…believes space is gay, because it has no technical orientation. 'One of the most annoying things about working in space is that there's no up and down,' she said, 'Because there's not a central gravity field, depending on where you are in space.' Most space 'orientations' are chosen out of convenience. For example, because the galaxy has a center, astronomers created a coordinate system based on a 'galactic north pole.' Human beings cannot possibly understand and make judgments about something without first putting it into systems they understand. Science, in that way, is inherently heterosexual, and space as it's written plainly in the stars, is queer" (Anderson 2020).

All space programs put a high premium on recruiting people to their astronaut, cosmonaut or taikonaut programs who are exceptionally physically fit—for Earth, that is. They need to be, it is argued, because of the profound strain and stresses of space flight. Therefore recruit the people who are best fit for Earth and send them to a place where humans don't fit in at all unless encapsulated in an elaborate bubble simulating the atmospheric features of Earth, whether it be a spaceship or a spacesuit.

Rose Evelein (2019) and Samantha Timmers (2020) present strong evidence that certain "disabilities" may make people more fit for the rigors and requirements of space than are so-called "able-bodied" people, especially in emergencies. When the lights go out, or smoke obscures vision, who is better able to continue functioning as usual than someone who already "seeing impaired"? While more arms (and a good prehensile tail) might be useful, long, sturdy legs are definitely a handicap in conditions of weightless. "Crip bodies were built for space travel. Crip minds already push the outer limits". Indeed, I long speculated that a lot of genetically-based "diseases" exist in order to be called into service by evolution when environmental conditions in Earth or not-Earth release them.

AstroAccess states:

> We are dedicated to advancing disability inclusion in space exploration, not just for the benefit of marginalized communities, but for the benefit of all humankind. This project is our first step in the direction of a new reality, where disabled astronauts not only exist but are trusted crew members, essential to the mission. When disabled people have equitable access to all jobs, taking on humanity's most complex tasks, perspectives change. Access to space changes the worldview not only of aspiring explorers, but of those that employ them, and most importantly, those that look up to them. By generating new opportunities for disabled

> scientists to succeed and by providing activists a platform, we have the power to inspire the next generation of scientists and world-changers. (Astroaccess 2020). https://astroaccess.org/about/

While people with all varieties of disabilites and enhancements have greatly contributed to the science, technology, and administration of space programs, some have also been enabled to experience weightless in parabolic airplane rides as a prelude to space, one of whom was Hayley Arcemeaux who also was aboard the SpaceX flight of 2021.

US millionaire Dennis Tito became the world's first space tourist on April 30, 2001, aboard the International Space Station. Subsequently, 250 non-cosmonauts/astronauts have visited the ISS. However, on July 11, 2021, Virgin CEO Richard Branson flew in suborbital space aboard a Virgin Galactic rocket plane. Nine days later, Amazon CEO Jeff Bezos' Blue Origin rocket spent 11 min also in suborbital flight on July 20, 2021. Virgin Galactic, Blue Origin and Elon Musk's Space X each flew their first tourist-focused flights in 2021, transporting several people with minimal training in professional spaceflight even though SpaceX's flight lasted three days in orbit. Indeed, one of the four, Hayley Arceneaux, had prothetic leg bones that would have prevented her from being an astronaut with NASA. Gay (and furry) Cameron Bess flew on Blue Origin. All private space companies may be reviewing their policies about who can qualify physically, mentally, ethically for space flight as they endeavor to boldly go.

Roscosmos (the Russian federal space agency) brought two sets of space tourists into space in 2021, including a mission with Space Adventures that has helped fly super-wealthy tourists to the ISS over the last 18 years. Russian actress Yulia Peresild, and producer-director Klim Shipenko spent 12 days in space filming a movie. The two civilians underwent rigorous training.

So my main point here is that all we know about actual humans in space comes within the context of very small groups of extremely unusual and atypical people. Initially, essentially all astronauts and cosmonauts were male test pilots, truly superhuman in every way. They were highly-disciplined though exuberant risk-takers who were willing to conform their thought and behavior to test the bounds of a largely unknown and completely new environment within spirit- and bone-crushing capsules. Even when opportunities for experiencing orbital space flight were opened to "ordinary" people, they have had to go through long periods of selection, training, and more training that often exceeded the time they actually spent in space (and many, so trained, never had a chance to fly at all). Many space tourists who claim the title of astronaut have been extremely rich, and few actually flew beyond the boundary between Earth and space, and back. So people who have flown in space so far are not like some random group of ordinary people who get on a scheduled airplane flight from Dayton to Daytona. To the contrary, they are extremely a-typical folks in every way.

But not only is the sample of humans from which we can make generalizations about human space flight very small and unrepresentative of humanity, but also the tools of social science which have been used to study and develop valid generalizations about human space flight are themselves comparatively new and problematic—scarcely older than the Space Age itself and ethnocentrically-limited. In contrast with the natural sciences which are more nearly globally-based, the conclusions of the social sciences came initially from studies of people (largely male) in western countries, Japan, and Korea. Even more seriously, while NASA now requires that all astronauts hold a master's degree in a STEM field, including engineering, biological science, physical science, computer science or mathematics, from an accredited institution, most of the space agencies and industries, and especially their sponsors and funders (meaning legislators and/or other political actors) have been extremely suspicious of social science, reluctant to fund or even allow properly-designed social science experiments or observations to be carried out as space activities. Thus, what we think we do know about human space flight comes—with some important but rare exceptions—from anecdotes or inadequately designed and carried out studies about people who have experienced the kind of paltry space flight available so far.

## 13.3 Earth Analogs for Not-Earth

Moreover, a very big source of information, or speculation, about space flight has come not from actual space flight but from analogs—from activities on Earth that are thought to be similar to those in space. So what are the conditions of space for which there are analogs we can study on Earth? Here is a typical list:

Population: small, homogeneous: all, or mostly, male; mono-cultural; same age-cohort; limited range of occupations.
Habitat: cramped and crowded; no privacy; visually sterile; constant noise—or utter silence; in constant motion; 24-h "day"; reduced or increased gravity; weightlessness; poor or dangerous air quality; variable air pressure; extreme heat or cold; dust or mold; high humidity; micro-organisms or insects.
External environment: very cold or hot; climate extremes; under water; underground; toxic atmosphere; solar or other radiation; asteroid bombardment; higher/lower gravity; higher/lower air pressure; all dark/all light; one day longer or shorter than 24 h.
Isolated from the rest of human society for extended periods.

And here are some places on Earth where humans are/have been that have some of these characteristics: remote small islands; rural farms; national parks/wilderness; prisons; oil rigs; sailing ships; *underwater*: divers, submarines, sea labs, aquanauts; *underground*: miners, cavers; early pacific island voyagers; early Western explorers; mountain dwellers and climbers; desert travelers and settlers; polar explorers and settlers; designed experimental habitat dwellers (e.g., isolation booths and tanks, various space simulators, Biosphere 2).

Indeed, are there any things in space without any earthly or human counterpart from which we can analogize?

*But wait!* Are we sure we have listed the most useful characteristics of space and the most useful Earth analogs? We may have fallen in the trap that the national space programs have set for us. We have imagined space settlements to be sterile, Spartan, military-like hutches in extremely hostile environments, which they might (or might not) be in the very initial period. But how many people will choose to go to space if it is eternally noisy, crowded, dangerous, and altogether unpleasant? We must not assume that just because military test pilots were the first persons in space, and because the first spacecrafts were so inhospitable (and still are), that future human space flight, and habitats, should, much less must, also be so rugged. Early spacecrafts—perhaps out of necessity, perhaps not—were designed according to criteria which assumed that abnormal humans, who were exhaustively conditioned to accept extremely hostile and inhumane conditions and environments, would function within them. It is curious that even now most actual designs are still extraordinarily cramped and stark, assuming that abnormal humans, for the honor of being designated an astronaut, can and will be made to fit within whatever is designed for them. The reason may be said to be economic necessity, but that seems a lame excuse: probably most people who inhabit space stations drive on Earth a private car that is roomier and more comfortable than their spacestation quarters. No, the reason cannot truly be that we are materially poor. Could it be instead that we are spiritually and esthetically impoverished, valuing machines and a kind of "economics" over humans?

Of course, the space entrepreuers who strive to lure rich people into paying for a few awesome and sickening moments in space know that they must give their clients a ride even more comfortable than that of their luxury auotomobile—or yacht—and they are doing that. That must become the norm.

So, the most useful Earth analog for space settlements may lie in one of the places listed above, and that is the remote small islands in the South Pacific. They are considered to be as close to earthly paradise as we can imagine. But, for example, Hawaii is one of the most remote spots on Earth and one of the last to be discovered and settled by humans. Until the invention of the jet airplane, only the rich and leisurely, or poor and outcast, could get there. After the waves of early Polynesian travelers arrived and settled in perhaps about one thousand years ago, Hawaii was an almost impossible place for most non-Polynesians to travel to, or to thrive in. Now, mass tourism is the number one, booming industry in Hawaii, and it took a lot of PR (as well as technological and economic change) to make it so. The Hawaii Tourist Bureau doesn't want anyone to know about the Dark Side of Paradise, and so it paints such a glowing picture that tens of thousands of people each year endure long, painful, cramped insulting airplane flights, and dump bucketsful of money into the economy simply because "Hawaii" has come to mean "Paradise" to most of the rest of the world, whatever it may mean for those who actually live there (it means "as close to paradise as I can get" to me).

That is to say, the Earth analogs we should be looking for should be those that express the very best aspirations for freedom and community we humans have ever imagined or dreamed of—and created. Not-Earth should be portrayed as "Hawaii", not as hostile wasteland where all must live as prisoners. But that is not the kind of analogs we have studied. Since the immediate future of human space flight may well continue to be one of spirit-crushing habitats in challenging, strange, and indeed hostile environments, Earth analogs to space so far have had those characteristics, and that is a big PR mistake. It might be acceptable for now, but if we really want "ordinary" people to go to Mars and elsewhere, we need to stop calling space "hostile" or "an extreme environment", and do what the Hawaii Tourist Bureau does—call it "a paradise" with great stupefying drinks, nubile women, and rugged, but sensitive, men while fully welcoming and embracing also all manner of LGBTQIA+, artilects, and cyborgs. Human perception of an experience (called the "definition of the situation") is in some ways more important than the objective "situation" itself. Some things, if approached with the expectation of being positive, are experienced positively, while the same situation, if defined negatively, may be experienced negatively. Moreover people's initial attitudes towards any environment typically change with prolonged exposure to it.

## 13.4 Scant Social Science About Humans in Space

Very little social science research has been carried out on actual space flights not only because funders don't like the social sciences, but also because the results of such studies have been regarded as being dangerous for any one person's future in space if their opinions or behaviors are revealed to be anything other that "normal". Candidates for space flight and astronauts in space resent, and indeed are fearful, of all tests that might reveal any deviance on their part. This is true of the medical tests they must necessarily endure, but even more so of any social science tests since they have faith in the former, but not the latter. Consequently, human thoughts and behavior are the least studied aspect of space studies (in terms of funding, number of studies, number of researchers, replications, etc.). Physical/geological and biological/environmental natural science aspects of space are the best studied while issues related to designing, constructing, operating and maintaining spacecraft are next. There have been very few properly designed, funded, executed, analyzed and published social psychological studies. The few that do exist are often those called "human factors" which seek to measure and improve human work productivity, and those related to organizing and managing a space activity in order to further the political or commercial interests of the sending/funding organization. Many excellent biological and physiological experiments useful for human space flight have been performed on the International Space Station, but I am not aware of any specifically-designed social science ones, in spite of the fact that the multinational, multicultural, and to some extent multidisciplinary nature of the crew makes it a perfect laboratory for all kinds of useful social science research.

It is also the case that America's early antipathy towards social science was not shared by every space agency. Even though the social sciences within the Soviet Union were quite restricted compared to those in the west during the Cold War, impressive research in some social science areas was done for the Soviet space program, perhaps because the Soviets were planning for and involved in multi-member, multi-cultural crews for somewhat longer-duration flights well before the Americans were. Social science projects, and even humanities activities, have sometimes been part of space projects but seldom the kind of research into class, gender, ethnicity, sexuality, and governance that a social scientist would like to do. For example, while there has been a great deal of speculation, we know nothing verifiable about sexuality in space—cis, gay, or solitary. We have had serious studies about the complexities of conception, pregnancy, fetal development, birth, and survival of a few laboratory animals in space with speculation about what the situation for humans might be, but I do not know of any studies commissioned by space agencies about human sexuality in any of the various niches of space. There have been lots of interesting and arousing anecdotes, about prolonged weightlessness as an effective replacement for Viagra, for example, and there are suggestions that typical gay sexual activities are somewhat easier to perform in weightlessness than are typical cis positions, but these tales are often contradictory and unreliable. We know far more about religious practices in space than we do about sex, and yet sexuality is about as fundamental a part of being human as is imaginable even though space agencies pretend it does not exist and indeed may have already been expressed in space in some guise. At the same time, satellite companies have made digital pornography cheap and abundant, while one viable business model of private space flights might involve sex tourism on space "love boats". Space entrepreneurs surely have talked about it and some may feature it sooner rather than later. Anecdotal evidence exists that the privateers have been approached with offers of filming pornography in space but there has been none, it seems. Unfortunately, the current space cowboys seem even more reluctant than the space agencies to spend precious time and money on vital social science research and training of any kind, and so I can almost guarantee we will see a significant space failure sometime in the near future because of human behavior of some untoward kind if the privateers especially don't wise up.

But maybe it doesn't even matter. What we call space programs now are focused on only a few, minute aspects of what we will need to know—and will learn—as we move on to serious activities on the Moon, Mars, asteroids, and the rest of the solar system including perhaps off-planets on entirely artificial environments in Lagrangian points. We currently are looking at mites on a dustball in a broom closet of the New Century Global Complex in China in comparison to what lies ahead. Moreover, we are examining the mites only through the lens of STEM perspectives even though the hard work in the natural and biological sciences is increasingly being done by algorithms. What is needed urgently is active experience in the humanities and social science—perhaps at the intersection of queer and critical race theories.

Without clinging to notions of essential identity by saying so, cultural differences do influence human interactions. What may be acceptable in one society may be forbidden in another. Different preferences in food preparation and eating, body odors, social distance and many more things sometimes result in conflict on Earth. For example, Kung, Mehinaku, Arabs, Javanese, Malays, and Japanese, in that order, seem to prefer much more human closeness and contact than do most Americans. Americans maintain a physical distance between themselves and others that seems "standoffish" to many, while Arab's desire for close contact (and smells) alarms most Americans who are constantly backing away as friendly Arabs futility advance. Social distancing as a result of Covid-19 was probably a boon to many Anglo-Saxons while almost impossible to practice by people from some other cultures. What Americans may consider to be necessary "privacy" is deep pathology for others. Attitudes towards drinking behavior and drunkenness; sexual comments and touching; how to be sick and respond to others who are sick; and many more behaviors are very different and very dynamic from one culture to another (including not only national culture, but also gendered, occupational, religious and other sub-cultures), but none of these important cultural differences have been studied sufficiently in space though they have very frequently been commented on anecdotally and unreliably. You fly on an American spacecraft, you surpress all your desires and act like you are from Kansas; on a Russian spacecraft, like you are from...Moscow?

Technology influences behavior in important (and barely studied) ways. In the old days, when movies were shown to people in Antarctic winter-over settlements via 16 mm projectors, everyone watched them together as a shared communal experience. With videotapes, and then the Internet, the experience became mostly private. Generations have learned how to stare down and peck away perpetually at their iPhones, never permitting even the slightest eye contact with other humans.

If, as we move from Earth's orbit to the Moon, Mars, asteroids and beyond, the mission goals, composition of crews, internal and external environments, role of robots, artificial intelligence and other autonomous entities, and many other things will present us challenges and opportunities unlike those we have experienced so far on Earth or not-Earth. We will need to prepare for novel situations that will be encountered by individuals, by small groups, and by overall organizations. Foremost among the latter will be opportunities for new forms and processes of governance that have seldom been imagined much less developed and used on Earth.

In my experience, the most provocative and challenging examples of how to help humanity prepare for long-range space flight and especially the possibility of alien encounters were the meetings of Contact!, an annual three-day convocation of anthropologists, science fiction writers, space scientists and enthusiast at various venues in California from 1983 at least until 2016. Contact! was created and maintained by Jim Funaro, an anthropologist at Cabrillo College in California. There were speeches, panel discussions, exhibits (some of many years' evolution), but the highlight of each conference was Contact!—the arrival of aliens with whom the rest of the participants were to figure out how to communicate—that being a specialty of some anthropologists, the plot of many sci fi stories, but the experience, it seems, of only a few of us in UFO reality.

One of the most frequently-mentioned human consequences of actual space flight so far is the so-called "Overview Effect" first identified by James White (1987) and greatly expanded and recently elaborated on by White (2019). Astronaut Ron Garan (2015) described his life-changing experiences in space and on Earth as did Astronaut Nicole Stott (2021).

The overview effect is a transformational change of values, attitudes and often behavior that some astronauts and cosmonauts have said they experience as a consequence of seeing Earth for the first time from afar, whole, beautiful, precious, and without any political or other artificial borders and boundaries. Certainly the astronauts who walked on the Moon (even more so the one who had to stay utterly isolated and alone in a hovering craft while others walked the Moon) were awe-struck by Earthrise and the glimpse of the tiny, frail, solitary Little Blue Marble that is Earth with its thin blue halo of atmosphere protecting life from cosmic death.

On the other hand, it is possible that passengers on trips to Mars, an asteroid, or other more distant objects will feel something quite the opposite as Earth recedes to a mere dot and disappears entirely before their new destination comes dramatically into view. It is likely that humans circling and then landing on Mars will experience something far beyond any "overview" effect as they realize humanity has finally gotten out of the cradle and begun to explore the neighborhood of the inner solar system, and eventually even beyond.

I suggested earlier that the decision by President Kennedy to land a man on the moon and return him safely to Earth was not made as it might more "naturally" have been made. A "normal" science and technology-driven space program would have followed a long, slow, scientifically-solid process. It might have begun with the careful observation and mapping of the Moon first from Earth via increasingly powerful techniques; then by a series of closer observations from Moon-circling unmanned spacecraft; then landing robotic rovers on the Moon that carefully surveyed all of the surface (and below) while performing a series of experiments, perhaps then robotically-preparing a base (or series of bases) where humans, cyborgs with their artilect companions would eventually arrive, set up shop, engage in more exploration and scientific experiments; finally evolving into a permanent and expanding settlement to which "ordinary people" cyborgs and artilects from Earth might migrate for ordinary or extraordinary purposes; and so on to exploring and eventually building settlements of humans, cyborgs, transhumans, artilects on Lagrangian points, perhaps asteroids, perhaps directly to Mars.

But no. Absolutely to the contrary, the Moon landing was not a process of such a "normal" scientific, technological, industrial and human evolution but rather a bold/rash/stupid political decision made entirely for purposes of inter-national ideological rivalry alone, with almost no proper scientific-technological-human preparation or justification. Very importantly, while technology did exist or was created that got the assigned job done on time, well, and at enormous expense, other technologies that could have made the process less costly and more sustainable—including more advanced robots, AIs/AEs, cyborgs, transhumans, and posthumans among other things—did not exist at the time. If currently-looming resource, ecological, and political limitations do not make it impossible, these technologies should eventually emerge, transforming Earth and enabling the steady exploration of the solar systems and cosmos.

When the ideological purpose was achieved with American landing on the Moon and the space race won and lost, long-range space programs everywhere came to an abrupt end. In the early 1970s, both the Soviets and the US ended their premature programs of long-range human spaceflight that many expected would continue (settlements on Mars by 1980!!). After a period of uncertainty very similar to the present, both the Soviets and the US (soon joined by Europe, Japan, and Canada) eventually settled on an uneasy balance between human space programs of Earth-orbiting spacecraft on the one hand, and robotic space programs of Earth-orbit-based space observation platforms and solar-system exploring and deep-space voyaging vehicles, on the other.

## 13.5 Space: Is it Just a Job?

Subsequent decades of work on Skylab, Salyut, Mir, and the International Space Station routinized certain space activities while atrophying, technologically and psychologically, the once impressive "space muscles" that were built-up during the Space Race era. While there have been many astronomical discoveries and important technological developments, as well as improvements in human/robot interaction during this time, from the point of view of most people not directly involved in the space industry, space seems to have become just a regular job, done off-Earth perhaps and so requiring big, expensive and cranky buses with dedicated bus drivers and passengers who go back and forth to work at a science lab that has rather spectacular views where routine science and technology-testing is performed. The space program has become just another high tech job done in a fairly remote place, but a place that is not psychologically more remote than it is, say, for someone who lives in Hawaii and has to fly to Europe and back frequently as part of her job. The generally safe and predictable nature of space station activities is, perhaps, one reason for relative lack of popular support for or interest in space now. It is just routine "going to work" for some people in the space industry. Nothing more. This conclusion is reinforced by the fact, mentioned before, that when the routine is broken, and disaster strikes (as in the catastrophic explosions of the Challenger and Discovery space shuttles), American's interest in spaceflight spikes sharply up, only to settle back down to its very low level when normality and predictability returns. It is hard to be excited about most astronaut's day job.

With the end of the American shuttle period, the future of American participation in human spaceflight activities became more uncertain than ever before. Like Apollo before it, the shuttle program was cancelled before a new American program was determined, much less underway. It was profoundly sad and demoralizing to walk through the gigantic empty buildings at Cape Kennedy in Florida, stripped clear of almost all remnants of its glorious past. What was it all for? Is this it? many wondered.

Currently there is hope that the American program will revive significantly and that even if it does not, that China, India, Russia, Japan, Europe, Iran, Brazil, Nigeria....other national space programs will be what is necessary to increase the human and posthuman presence in space, even if America's presence declines.

At the present time, many eyes are focusing on the space entrepreneurs who have had spectacular successes in the tasks they have set for themselves, and promise much more.

"Those recent and upcoming missions have created the perception that space tourism has finally arrived—maybe a decade and a half late.... Virgin Galactic has more than 600 customers and reopened ticket sales in August, while Bezos said in July the company had as a backlog of nearly $100 million, although hasn't disclosed how many people have signed up. After Inspiration4 splashed down, a SpaceX executive indicated there was enough demand for orbital flights to support five to six Crew Dragon missions a year" (Foust 2021a). "It's a small sign of a growing maturation of the commercial human spaceflight industry. After years—decades, really—of waiting for it to arrive,...three companies flew private human missions from the edge of space to altitudes above the space station in 2021. Private human spaceflight become more regular, but not routine" (Foust 2021b).

"Space 4.0 represents the evolution of the space sector into a new era, characterised by a new playing field. This era is unfolding through interaction between governments, private sector, society and politics. Space 4.0 is analogous to, and is intertwined with, Industry 4.0, which is considered as the unfolding fourth industrial revolution of manufacturing and services" (ESA 2016).

Perhaps so, but I am profoundly uneasy with the way the privateers are undertaking space ventures now. First of all, the "fourth industrial revolution" is a phrase very popular in Europe, Japan, and Korea. I have frequently spoken out against its use. It treats the transformational potential of developments in electronics, robotics, AI, ALife, biotech, neuroscience, nanotech, and material sciences as just one more phase in the old industrial revolution even when elaborations of the term suggests it is much, much more: "The First Industrial Revolution used water and steam power to mechanize production. The Second used electric power to create mass production. The Third used electronics and information technology to automate production. Now a Fourth Industrial Revolution is building on the Third, the digital revolution that has been occurring since the middle of the last century. It is characterized by a fusion of technologies that is blurring the lines between the physical, digital, and biological spheres" (Schwab 2016).

However, I share the sentiments about this unfolding moment in space activities, but not necessarily all the reasons expressed by Layla Martin: "The American public is reassured that space is an arena fostering unity and cooperation. Why is that not what we are seeing?...We are being sold the idea of space, in that feel-good It's a Small World kind of way but are witnessing a handful of rich guys jostling for position like bloodthirsty contenders pre-IPO" (Martin 2021).

Yes, but it is not unusual historically for the functional equivalent of billionaires to lead the way in social and technological developments. As I have repeatedly stressed, marginals and misfits have always played important roles in space and continue to do so. Billionaries are just weirdos of a different sort and we should press them to use their talents more for the common good and less for their private greed. I am especially worried about and have expressed my displeasure at their blatant disregard of the environmental consequences of their actions, especially when they move from tourism to mining and actually establishing "colonies" on the Moon, Mars, asteroids, and elsewhere.

I am perpetually vexed by the willing thoughtlessness of humans to "throw away" our waste—out the window of our speeding car, into a landfill, or in the case of the space industry, into the "vastness" of space in the belief that it will eventually be burned up as gravity pulls it back to Earth, and that it will not endanger other space objects until then. The consequences of this very unethical though extremely common (and often economically-justified) behavior is now so serious that current and future space activities are increasingly endangered and, unless our junk is cleaned up and further pollution prevented, may render some future space activities impossible. I have heard for thirty years promises not to keep proliferating orbital debris and schemes for cleaning up what exists, but I am still waiting for real and effective action to begin, as we continue to spew out more garbage. I am not convinced space cowboys are likely to be more conscientious in this regard than are national programs.

Similarly, the disdainful disregard for the destruction of all forms of life on Earth almost certainly will extend into space activities, especially those that are commercially driven. "Life" is a very difficult thing to define unambiguously, but we are not now making serious, adequate preparations for preventing harm to life we might encounter in space—other than perhaps protecting ourselves from possible contamination from them! Fossilized life forms which might provide "useful" information for humans as well should receive ethical considerations on their own but are viewed as just rocks to be crushed at will.

But what about rocks themselves? Shouldn't rocks, landscapes, and even abiotic natural formations be left alone? Landscapes, rivers, mountains, and other features should receive respect and reciprocal considerations from humans. Indeed, entire biospheres should be cherished as they exist. There should be limitations to what humans can do to a biosphere and its parts. Our failure so far to grant respect and deference to trees and rocks may be leading to the extinction of humans on Earth whose existence is dependent on the services of abiotic as well as biotic forms. There may be reciprocity at a planetary scale going on now on Earth. We have done it unto nature so nature is doing it back unto us. We should not extend our ignorance and arrogance willfully into space in the name of profit and prestige.

While we may have already done irreparable damage by space intrusions, humans should not land on the surface of any object in space until we know a whole lot more about what "life" and "intelligence" is, and their presence in space. We should not "boldly go", but go cautiously, sensitively, and ethically into space. First People are said to ask the permission of a deer before they kill it for food and materials, and many people "say grace" before eating anything, thanking God and nature for providing the sustenance humans need. Many Hawaiians ask permission of a rock before they remove it for their use. So also might humans humbly ask permission of Mars and the asteroids—and wait for an affirmative answer—before terraforming or exploiting them, if we should exploit them at all.

Some time ago, Christopher Stone, the son of the rabble-rousing journalist, I. F. Stone, wrote an article for a relatively obscure law review that quickly became the subject of a Supreme Court dissent and a book of considerable impact, titled,

"Should trees have standing? Towards legal rights for natural objects" (Stone 1972). "Standing" is a legal status that gives one the legal right to have their case heard in a court of law. Since that time, the matter has been a languishing emerging issue, taking a long time to wend its way up the "S" curve of emergence to becoming what seems to be a trend and soon, perhaps, a full-fledged problem/opportunity:

> In 2017, an exceptional incident occurred. Whanganui River became the first waterway in the world to get legal personhood. The third-longest river in New Zealand can now be represented in court and has two guardians to speak on its behalf. Environmentalists and Indigenous rights advocates praised this unprecedented event. Other countries have also followed the fascinating Whanganui's example. Two rivers in India have been declared legal entities, and Bangladesh gave all its rivers legal rights (European Wilderness Society 2022).

And now a lake in central Florida is demanding its rights as well (Kolbert 2022).

What about artificial intelligences and environments? Consideration for the rights of robots was argued early on by two researchers in the planning department of the Hawaii State Judiciary (McNally and Inayatullah 1988) and the matter is picking up steam now as well.

On the other hand, many years ago, Martin Krieger wrote an article that startled me, titled, "What's wrong with plastic trees?" What startled me was not the title or the notion—it was a completely reasonable question, it seemed to me—but that it appeared in *Science* magazine, the official publication of the American Association for the Advancement of Science of which I am a member. One is not likely to find such an article in *Science* now. It is much more cautious and staid, but there were in fact other thought-provoking cutting-edge articles in *Science* in that era.

Krieger prefaced his article with a well-known statement attributed to Ronald Reagan when Reagan was a candidate for governor of California: "A tree's a tree. How many more [redwoods] do you need to look at? If you've seen one, you've seen them all." Krieger took that outrageous statement and ran with it, concluding,

> What's wrong with plastic trees?
>
> My guess is that there is very little wrong with them. Much more can be done with plastic trees and the like to give most people the feeling that they are experiencing nature. We will have to realize that the way in which we experience nature is conditioned by our society—which more and more is seen to be receptive to responsible interventions. (Krieger 1973: 453)

Now the issue should be framed more broadly as "what's wrong with artificial intelligence, life, and environments?" And the answer is nothing, except that they are still fighting for the right to their identities and for others to respect it, like so many others.

But in truth, none of this should be framed as "rights". That is an essentialist, pugilistic way of thinking. These are instances of *shu* (恕)—the mutual interactions and responsibilities between robots, human beings, human becomings, transhumans, posthumans, and all entities and environments whether natural or artificial.

Some people may argue that this is ridiculous; that Earth and the universe were made for humans to exploit and we should do so without the slightest qualms of conscience: "God blessed them and said to them, 'Be fruitful and increase in number;

fill the earth and subdue it. Rule over the fish in the sea and the birds in the sky and over every living creature that moves on the ground'"" [Genesis 1:28]. Will humans be deterred from our desire to exploit space for economic, military, nationalistic, or other purposes while fretting about empathy for rocks, trees or robots? Highly unlikely. The entire history, present, and a probable future of junk on Earth as well as in space strongly suggests that humans at best only feel twinges of regret over the consequences of stupid acts. We seem incapable of exercising ethical foresight before we act regrettably once again.

As we render all of Earth artificial and synthetic, we surely will extend the Anthropocene Epoch into space—if we are able to move into space at all. I am convinced that most humans vastly underestimate the speed and magnitude of disruption that the forces of climate change are wreaking and will increasingly provoke. It is not fanciful to wonder if we will have to abandon not only "optional" activities such as those related to space and big science, but also even those that seem most central now—the design and manufacturing of endless new consumer goods and killing weapons—will end. We will focus instead on basic issues of survival—the rising and intruding seas, the droughts and floods as climate becomes completely unpredictable, and famine widespread.

Opposition to all space programs is widespread. In addition to anguish over the reckless expansion of militarism and environmental destruction, space programs divert money and talent desperately needed on Earth so that a privileged few might enjoy unusual views and some moments in weightless and/or the reduced gravity of spaceflight:

> A rat done bit my sister Nell.
> (with Whitey on the moon)
> Her face and arm began to swell.
> (but Whitey's on the moon)
> Was all that money I made last year
> (for Whitey on the moon?)
> How come there ain't no money here?
> (Hmm! Whitey's on the moon)
> Y'know I just about had my fill
> (of Whitey on the moon)
> I think I'll send these doctor bills,
> Airmail special
> (to Whitey on the moon) (Scott-Heron 1970).

These are the closing lines to a powerful poem, "Whitey on the Moon", that Gil Scott-Heron wrote, delivered (with bongo drum accompaniment), and recorded in 1970 on a phonograph record titled *Small Talk at 125th and Lennox* by Atlantic Records. It was a tremendous hit among Black and other communities, and has reverberated over and over again down to the present in cultural critiques of various guises. Scott-Heron's equally powerful and popular poem, "The Revolution Will Not Be Televised", also appeared on the same record. Both, and the record over all, can be considered intimations of Afrofuturism, now in full flower. Grace Gipson quotes,

"Afrofuturist writer and artist Ytasha Womack [who] states, 'Afrofuturism is an intersection of imagination, technology, the future, and liberation'. The intersection that Womack describes is not solely the domain of artists, musicians, authors, and scholars; in the 2010s, with digital technologies, numerous people are identifying as Afrofuturists and are using Afrofuturist ideas and concepts to educate, share stories, fight oppression, and help build communities in need" (Gipson 2019: 84).

In "Afrofuturism envisions space in 2051", Russell Contreras writes, "Black science fiction writers and artists known as Afrofuturists say the next 30 years of space exploration could address legacies of racial terror on Earth if people of color join ventures and help reimagine human life among the planets" "No Black astronauts went to the Moon. That sparked spoken-word artist Gil Scott Heron to record his iconic poem, 'Whitey On The Moon', in 1970, tackling the hypocrisy of white space exploration without Black people" (Contreras 2021).

More specifically, Philip Butler in "The Black Posthuman Transformer: A Secularized Technorganic" envisions "...the Black posthuman transformer is the complete merger of the human entity with nature (reconfigured as machine) at the cellular level. It is a technorganic entity (meaning its extra-human technology is infused at the cellular levels) grounded in human biotechnology." "As a futuristic entity, it is a gender shifting complex autopoietic system of infinitely augmentable capability. Already the Black posthuman transformer is able to account for its own entanglement with biology, gender, sexuality, culture, identity, art, history, technology, futures, and society, simultaneously" (Butler 2019: 64).

## 13.6 Adapting Humans for Not-Earth

Thousands of reports over hundreds of years have made it clear that the human body evolved to fit the environment of the surface of Earth during the late Pleistocene/early Holocene Epoch. It is dangerous and difficult for humans to go too high—even to the mountain tops of Earth, and certainly any higher, or to venture too low, especially underwater, without being encased in a bubble of some kind that simulates the essential elements and as many of the desirable aspects of Earth's surface conditions as possible. Return from these locations to the surface of the Earth needs to be done carefully so as not to shock the body into unhealthy reactions. Prolonged exposure to heights and depths provokes adaptations that must be carefully revoked in the process of returning to surface conditions.

Space scientists have known for a very long time from numerous studies that humans are absolutely not fit for the nearby environments of Earth—Moon, Venus, Mars and interior asteroids. The only way humans can survive is by keeping them tightly inside capsules and spacesuits. But those are still dangerous, not only because of the technical challenges of escaping Earth's gravity itself but also the tremors of liftoff itself and passage through various transition zones. While space agencies have suffered the death of astronauts during faulty liftoffs and landings, and while spaceships and spacewalkers are constantly being bombarded by microscopic space debris, no one has been killed while walking in space nor has a spaceship been fatally damaged or destroyed by any collision with debris in space so far. However, Apollo astronauts in 1972 were fortunate not to have experienced a "solar energetic particle event" while walking on the Moon since it was during the height of a solar cycle when solar flares and particle events are most abundant. If humanity's first moonwalkers had been fried on the Moon, the course of history might have been altered during this competitive Cold War period.

Simply being in space in the currently-available protective environments is hazardous for the healthy performance of all bodily functions. The longer away, the longer it takes to return to "normal" when back on Earth, with some damage irreparable. The longest single human space flight was by the Russian Valeri Polyakov who circled the Earth for 438 days in 1995 in the Mir space station. Most flights have been only of 10–20 days duration though people tend now to stay on the International Space Station for about six months or longer. So we really don't know much for certain about the effects of long-term stays on the Moon or long-term trips and stays on Mars.

NASA recently took advantage of the fact that one of its astronauts was going to spend almost a year on the ISS while his twin brother would stay on Earth to conduct a comparative study of two genetically similar humans, one in orbit, one on the surface of the Earth—a biologist's (not to mention Einstein's) experimental dream come true. A detailed report of the study stated: "Longitudinal assessments identified spaceflight-specific changes, including decreased body mass, telomere elongation, genome instability, carotid artery distension and increased intimamedia thickness, altered ocular structure, transcriptional and metabolic changes, DNA methylation changes in immune and oxidative stress—related pathways, gastrointestinal microbiota alterations, and some cognitive decline postflight" (Garrett-Bakelman et al. 2019).

Data from one year on the ISS is valuable but grossly insufficient for longer durations on the Moon and Mars, especially for those for whom the Mars trip will be one-way.

So far space programs have insisted that the only solution is to build habitats in space that are as Earth-like as possible. There have been many suggestions about how this should be done. The subject of space architecture is fascinating, and many ways designs for space may help us address the many novelties that the Anthropocene Epoch presents to habitats on Earth. No matter how "Earth-like" environments on the Moon and Mars may be, they will not be Earth, and so evolutionary processes will begin to work again on humans even in the most hermetically-sealed habitat in space. *Homosapiens, sapiens will* become something else—"naturally"—whether we like it or not, or think it is "ethical" or not, or destructive of our identities as human beings, or not. If any of us leaves Earth to live elsewhere for an extended period of time we will become, and no longer merely be. The most fundamental identity for most of us now will be gone, along with all the fears and fantasies about "the extinction of humans" and the loss of our precious heritage. To be sure, the first experiences for humans of relatively extended stays off-Earth will probably be within highly controlled environments simply because we don't know how to redesign and engineer posthumans for space yet. But what was once a taboo—ridiculous—sacrilegious—subject is finally being discussed seriously in a few places. Barring the loss of the processes that make contemporary science and technology possible, some of us *will* continue to become posthumans and move into Not-Earth. Others will follow.

> As George Church, a Harvard geneticist and leading synthetic biologist, argues: "One likely path for risk reduction in space does seem to involve biological engineering of adult would-be astronauts." He has identified 40-some genes that might be advantageous for long-term spaceflight (and would benefit those who stayed behind, too). His list includes CTNNBI, which confers radiation resistance, LRP5, which builds adamantine bones, ESPA1 (common in Tibetans), which allows people to live with less oxygen, as well as a host of genes that might make us smarter, more memorious, or less anxious. The menu even includes a gene, ABC11, which endows its possessors with 'low-odor production,' a friendly trait in a confined space. (A spaceship with standard humans smells like the Harris County Jail, according to one recent inhabitant of the space station.)
>
> Church cofounded Harvard Medical School's Consortium for Space Genetics, along with other prominent biologists like the anti-aging researcher David Sinclair, in order to study human health in space and promote exploration. He imagines "virus-delivered gene therapies, or microbiome or epigenome therapies" that astronauts would take to transform their biologies. "Quite a bit is already known about resistance to radiation, osteoporosis, cancer, and senescence in mice," he says. Church stresses that many of these genes are already targeted by pharmaceutical companies, with drugs in clinical trials. Using gene therapies as a kind of preventative medicine for astronauts isn't so far- fetched (Pontin 2018).

Chris Mason, a geneticist and associate professor of physiology and biophysics at Weill Cornell University in New York, was a principal author on the NASA Twins Project. He is also among those exploring the genetic structure that allows certain forms of life on Earth to thrive in environments that humans find "extreme"—heat, cold, motion, radiation—indeed all the features of some niches of space. Some of those genes exist in humans and might be able to be enhanced and mobilized for space. Others might require CRISPR or other forms of synthetic gene therapy.

Mason and a group of researchers have created the Mason Lab for Integrative Functional Genomics. Their website states:

> The Mason laboratory is working on a ten-phase, 500-year plan for the survival of the human species on Earth, in space, and on other planets. To that end, we develop and deploy new biochemical and computational methods in functional genomics to elucidate the genetic basis of human disease and human physiology. We focus on novel techniques in next-generation sequencing and algorithms for tumor evolution, genome evolution, DNA and RNA modifications, and genome/epigenome engineering. We work closely with NIST/FDA to build international standards for these methods and ensure clinical-quality genome measurements and editing. We also collaborate with NASA to build integrated molecular portraits of genomes, epigenomes, transcriptomes, and metagenomes for astronauts, which help establish molecular foundations and genetic defenses for long-term human space travel (The Mason Lab 2020).

Five hundred years! Now those are my kind of futurists, sort of. Actually, they make some assumptions that I don't necessarily support. Their "Case for Existence" justifying the project states:

1. Assume that humans are the only species or entity with self-awareness of its own extinction.

2. Assume that existence is essential for any other goal/idea to be accomplished ("existence precedes essence").
3. Therefore, humans who wish to accomplish any goal/idea should ensure the existence of our species, and all other species that enable our survival.

I see no basis for assuming the first statement and depending on what they mean by the "existence" of our species (humans) I may or may not support the second and third. I am interested in *homosapiens, sapiens* continuing to evolve, to become, to shed or modify our current physical and social containers as necessary so we can spread across the cosmos almost certainly not as anything that looks or acts like "humans" now but who are nonetheless our direct genetic/prosthetic descendants.

Of course Mason is not alone in making these kinds of assumption. Humans are indeed spooked about their inevitable individual death and have invented all sorts of death-defying beliefs and rituals, such as religion, to deny what is real. Given our mix of killing and loving propensities, even the Judeo-Christian God can be forgiven for contemplating putting us out of his misery, ending it all, and starting the experiment all over again, perhaps by tweaking up the care and empathy genes and toning down the greed and violence ones. Having been fortunate to have a granddaughter with Downs Syndrome, I am fully convinced that a planet full of people so gifted would make Earth as close to a paradise for all as could be imagined.

Recently there have been several expressions of concern about the extinction of humans, perhaps starting with Bill Joy's startling 2000 essay in *Wired*, "Why the future doesn't need us" (Joy 2000), and including Phil Torres' forthcoming book: *Human Extinction: A History of Thinking About the End of the World* (Torres Forthcoming). Indeed many of the most popular arguments for long-term space flights and human bioengineering are about preserving humanity and living forever, sentiments I do not share. But if such arguments, along with military might, national and personal prestige, financial gain, sexual adventures, religious missions, and scientific knowledge will rock some of humanity out of our cradle and into space, then let it be...uh, become.

Especially impressive and squarely on point from my perspective is the work of Konrad Szocik and colleagues. Like Church and Mason, they have collected a team of experts who assume that the human body as now typically manifest must, can, and will be biologically modified if some humans are to move into not-Earth for long term exploration or settlement. They appreciate the role that cocooning biospheres and prosthetics play, but they focus on what is being done now, probably can be done soon, and what might be done later in specific biological modifications on various bases. They have produced an impressive number of monographs in scholarly journals as well as a comprehensive book dealing with specific technical and ethical issues surrounding the issue with contributions from twenty-three authors from around the world (Szocik 2020).

From my point of view, the case for and feasibility of human modification for space has been made in these recent ongoing research and development activities. Ways to do so have been identified clearly and plausibly. If steps are not taken rapidly and surely now, they eventually they will be. If not by Americans, then by other people, governments, privateers or venturers. It is no longer just speculation and fantasy and hope alone.

## 13.7 Cyborgs, Artilects, and Intelligent Environments

The human modification folks are currently making great contributions towards getting *homosapiens, sapiens* into not-Earth in something like their present bodies. Their contributions are an essential aspect of the entire venture, but they seem to rely basically on biology and I don't see much evidence that most of the cosmos is as enamored of biology as we humans are. We need to merge biological and nonbiological materials and processes effectively, as well as merging individual bodies into functioning interactive intelligent collectives and indeed environments.

Cyborgs are clearly an early part of that process—modifying the human body—or any organism—prosthetically. This is absolutely nothing new. As I have pointed out before, any extension of our body and central nervous system prosthetically makes us a cybernetic organism. We became cyborgs when we picked up a stick and began digging grubs out of a log with it. However, the term is more appropriate when we attach protheses to our bodies more or less permanently, such as by wearing clothes, or shoes, or glasses, or jewelry. It is even more appropriate when we literally fuse physical technologies into our bodies with dentures, pace-makers, hip-replacements, heart transplants, as well as computer chips, artificial limbs and on into the farthest reaches of biohackers now and in the futures.

As I keep repeating, I am focusing on what I observe humans—individual humans or communities—do as they go about their daily lives. I am not talking about any governmental program of planned human augmentation, though that of course would just be a more radical example of what I observe humans (and many other organisms) doing on their own from the beginning of time. So it seems likely that as space programs move beyond cautious and conservative national space agencies we might see private individuals and groups merging physical technologies with biological technologies if it will enable them to move from Earth into some of the myriad niches of not-Earth. Once we do move beyond our reliance on cocooned *homosapiens, sapiens* only—or even biologically-augmented *homosapiens* and/or artilects only—and enable a wide range of merged and novel posthuman entities—with all the attendant failures, tragedies, setbacks, objections, and successes that such processes will entail—we will be able to find comfortable homes in not-Earth.

The focus here so far on the metamorphosis of Earthkind into Spacekind (or Not-Earthkind) has largely been on the behavior of individuals or groups of individuals; of individual autonomous selves willfully acting and interacting with other individual autonomous selves as well as with groups composed of such autonomous selves. But interesting developments are occurring in living artificial materials, environments, processes and systems.

Among the many binaries being shattered now, the distinction between life and not-life is becoming murkier and murkier. Materials are becoming more and more lively—responsive, interactive, and indeed anticipatory wherever materials are used—in clothing, housing, entire urban conglomerations, indeed everything, even governance,

Yes, governance.

As I have said, I have had a long interest in new governance design. No social institution is as obsolete and as dangerous than are all current systems of governance, that of the United States of America especially (Dator 2006). They are all based on 17th Century ideas and 18th Century technologies with a few recent communication technologies thinly scattered on top. I have long taught that governance should be redesigned on the basis of cutting-edge cosmologies and technologies, and that not only governing structures and processes but also concepts and philosophies need to be similarly rethought and incorporated into new governance designs (Dator 2020).

Among the old ideas that need to be reconsidered, reformulated, or rejected is the concept of "rights" and the content of rights. The Confucian concept of shu 恕 (interrelationships and interdependence) that Roger Ames explained earlier in this monograph when we first began discussing "Human Becomings" seems much more vital now. Moreover, many Americans misunderstand the initial intent of the so-called Bill of Rights that were added to the original constitution as amendments. Many of the ten amendments were originally intended to forbid Congress from doing things the States thought they should be able to do. The portion of the First Amendment concerning the establishment of religion was not primarily intended to guarantee individual citizens the right to worship what and as they pleased. Rather an original intent was to prohibit the federal govenment from establishing a national church so that the states (some of which had official churches) could continue to have them. Over the years, for various reasons, the Bill of Rights have been reinterpreted by the US Supreme Court to mean things they almost certainly did not originally mean, at least to some who supported them. This is especially true of the Second Amendment concerning Arms (guns). More importantly there are many vital features of life in the Anthropocene Epoch that were unknown or trivial when the Bill of Rights were adopted that probably should be in such a Bill now.

One of the myriad areas in which saveguards against governmental power is needed, I pointed out in the early 1970s, is over the programing of computers that conduct governmental business, such as counting ballots and the like. This is a big issue now, but not in 1970 when I said that the software program and hardware configuration of any computer used to make or assist in making governmental decisions must be easily publically knowable and subject to public criticism, improvement, or rejection. I did not use the term "algorithm" then, but that would be something I would say now. Algorithms are being used to make unfathomably numerous decisions now in all aspects of our lives without our knowledge, consent, input, or influence, and yet the public does not have the "right" to understand what they are in order to assess their quality, accuracy, fairness, and impacts. I believed that of the simple computers of 1970. It is vastly more important now.

And I certainly believe it to be the case for the following proposal. A number of years ago, I included an article by Marcella Bullinga in the packet of reading for students in my governance design classes. Bullinga perseptively wrote:

> In the years ahead, technology will provide government and society at large with tools for a safer world and for automatic law enforcement. Permits and licenses will be embedded in smart cars, trains, buildings, doors, and devices. Laws will automatically download and distribute themselves into objects in our physical environment, and everything will regularly be updated, just as software is now automatically updated in your desktop computer.
>
> Innovations in government will enable us to have a safer environment for law-abiding citizens because built-in intelligence in our environment will minimize fraud, global crime, pandemic diseases, accidents, and disasters. Law-abiding citizens will gain privacy, while criminals will lose it.
>
> Making rules and enforcing them are important government tasks. Right now, laws are written down on paper and enforced by individuals. In the future, all rules and laws will be incorporated into expert systems and chips embedded in cars, appliances, doors, and buildings—that is, our physical environment. No longer will police officers and other government personnel be the only law enforcement. Our physical environment will enforce the law as well (Bullinga (2004).

The reaction of my students ranged from being horrified to being intrigued. Some clearly warned that this was another onramp to dictatorship, and the slew of books in subsequent years about surveillance capitalism and governance have identified, amplified and justified their concern. The solution in my view is not to ban these developments but to design "bills of rights" and other safeguards into them. This has not happened, in part because of the economic (as well as power) interests the owners of such algorithms have in keeping them secret. Economics and power should not be allowed to prevail here.

At the same time, Bullinga proved to be right about the phenomenon, its ubiquity, increase and irresistibility. For one thing, which Bullinga did not mention, asking humans to do what police often have to do is inhumane. Few lives are as stressful, dangerous, and unappreciated as those of police who frequently must do brutal things, including killing, to brutal people in terrifying, life-denying situations. How liberating it would be if we could replace humans in as many of those situations as possible, leaving it to intelligent environments to make it almost impossible for any one to do certain illegal things. And of course, *law-making* should still remain (or finally be placed) democratically in the will of The People via carefully designed, checked, and balanced (but not stalemated) electronic processes of governance, while removing as much of *law enforcement and adjudication* as possible from the whims, prejudices and anxieties of human beings seems like a great idea however difficult it might be to obtain. But it can be achieved by diligence, trial and error, and will.

And if not by processes of designed democratic participation in policy-making, then it will likely be achieved by tricks of capitalism, marketing, deception, lies and greed, as is happening now because, as long as electricity flows and data can be transfered "freely", most people seem prepared to give up their freedom and privacy in order to buy dodads or share intimacies cheaply and effortlessly.

So, as the technology facilitating smart everything continues to spread, our environments will be more and more under their influence, and humans will learn how to adapt and thrive within these new conditions just as we did in the past—or not, just as others did not in the past.

But here we have fallen into another trap. Bullinga is using intelligent environments to control. How much better for governance to be focused on enabling and enhancing desired behavior as much as possible, and not primarily on prohibiting bad behavior. Indeed much of the work on intelligent environments is focused positively. Augusto, et al. state in their "Intelligent Environments Manifesto":

"In order to help characterizing what we interpret by Intelligent Environments we list below some key principles we believe every Intelligent Environment should aspire to have:

P1) to be intelligent to recognize a situation where it can help.

P2) to be sensible to recognize when it is allowed to offer help.

P3) to deliver help according to the needs and preferences of those which is helping.

P4) to achieve its goals without demanding from the user/s technical knowledge to benefit from its help.

P5) to preserve privacy of the user/s.

P6) to prioritize safety of the user/s at all times.

P7) to have autonomous behaviour.

P8) to be able to operate without forcing changes on the look and feel of the environment or on the normal routines of the environment inhabitants.

P9) to adhere to the principle that the user is in command and the computer obeys, and not vice versa" (Augusto et al. 2013: 4).

This list is very similar to many of the key features of the steps in human—computer interaction that I discussed earlier, only Augusto, et al., stopped—for personal, political or ethical reasons, I suppose—before the steps where computers completely relieve humans of the burdens, you might say, of thought, planning, and ultimate control.

The point is that intelligent environments—both physical and virtual—are proceeding apace, and when combined with the other developments in AI, Alife, human modification, cyborgs and the rest, are propelling life and nonlife into realms and relationships they have never known before.

Which includes also the eventual merger of some living and nonliving symbiots into a hive of melded selves and wills, real and virtual, beyond individual identity.

## 13.8 Neither a Utopia Nor a Dystopa. Just a Topia

This mutation from Earth through the cosmos is a possible, plausible, and preferable future. But I do not claim that it will not be a utopia. It will not be a perfect world. It, and achieving it, will be as full of pain, anguish, cruelty and love as have all the worlds humans have lived in over the centuries. We absolutely must understand that striving to make everything perfect will inevitably and quickly produce a dystopia. The processes I have discussed will by no means lead to one inevitable future. What I have gestured towards is possible, plausible, and—to me—preferable. It is not a future that will or should result from governmental fiat, whether that government be somehow "democratic" or a religious, ideological dictatorship. It should emerge from the acts of competing, cooperating, complaining human individuals and groups. It will not unfold like toothpaste on the brush. We will experience starts and stops, failures and triumphs, and quarrels over which is which.

It is probably obvious that my thinking about the posthuman future has been influenced by thinkers like Teilhard de Chardin and his concept of the Noosphere; by Buckminster Fuller from whom I learned directly on several occasions as well as his numerous books; by Marshall McLuhan to whom I have referred so frequently; by Arthur C. Clarke (not his science fiction); by Ray Kurzweil, initially his *Age of Spiritual Machines* (1999) and then *The Singularity is Near* (2005); by Nick Bostrom and his book on *Human Enhancement* (2009) and others from the Oxford school. But I was never drawn to their evangelical certainty that their ideas pointed the way to human improvement and perhaps perfection, and the inevitability of their visions. I was more impressed by John Platt's "Acceleration of Evolution" (Platt 1981) and especially by Susantha Goonatilake who I also knew well, encouraging him to write *Merged Evolution: Long Term Implications of Biotechnology and Information Technology* (Goonatilake 1999) before Kurzweil popularized similar ideas.

Kurzweil and others are also morbidly focused on death, or rather on not dying; on living forever. They want new technologies that will treat death as a curable disease. Aubrey de Grey believes we can keep our bodies going the same way we keep old cars or houses going—by replacing aging parts with new parts before they wear out. Not me. I want to die. I think my living forever, no matter how heathily, would be horrible for me and others. As the timeline I summarized some pages above makes clear, bisexual reproduction is a relatively recent invention that brought perpetual novelty to Earth for the first time. And it also brought individual death. Crystals extrude and "live" forever. So do ameoba. But biotic beings die.

That is, the individual dies but life goes on—through biological offspring and through all the people one interacts with in her lifetime. As a parent of four and a mentor of thousands, I am quite satisfied with that. To live forever in my body and mind is to become a kind of crystal again. Instead, "I" came from nowhere. My consciousness of self emerged, grew, changed, and guided me as I lived among the many lives of Earth, as Ariel Yelen said in her eloquent poem quoted pages before. I near death (I devoutly hope!), and when I die that is it. That is why one should do her "best" while living, but for no other purpose than that, and without being a self-righteous and selfish jerk about it.

> Sleep softly . . . eagle forgotten . . . under the stone,
> Time has its way with you there and the clay has its own.
> Sleep on, O brave-hearted, O wise man that kindled the flame—

To live in mankind is far more than to live in a name,
To live in mankind, far, far more . . . than to live in a name.

From Vachel Lindsay, "The Eagle That Is Forgotten" (Lindsay 1923).

On the other hand, what de Grey proposes might be right down my alley, if he takes his auto parts replacement example seriously, and moves away from replacing worn out body parts with new bio-based ones; if, that is, he enables humans to move beyond biomodification to cyborganization to various kinds of *homowhateverus* as Ian Pearson labels them.

Some years ago, Manuel Delanda said that we humans "might just be insects pollinating machines that do not happen to have their own reproductive organs right now." (Davis 1992). I think we need to view the issue of our pollinations as we should view our children now: they are not smaller versions of ourselves—of me and a significant other—they exist because of our actions (and probably not intentions), and we are stuck with taking care of them as they mature. But they are never "us". They are themselves and we must celebrate that and whatever they continue to become. So also with our technologically-transformed successors. They are our children, so nourish, support, and love them now because you can't leave them and the scars on your memory.

Importantly, it is not just humans, posthumans and artilects who will be involved in this transformation. The Anthropocene means what it says: it is the age where human acts influence everything else, including the biosphere, for better or worse. But who is to say what is "better" and what "worse"? Every new technology in the past has helped redefine what it means to be "human" by making it possible to do things we couldn't do before and making unnecessary things that had been essential before. We cannot judge the value of new technologies until we have lived with them and see how they change us—almost certainly in anticipable but not predictable ways. And once we have fully engaged new technology, it is very, very difficult, probably impossible, to go back to the old ways again. Technology is neither neutral, demonic, nor transformative. But it is mutative. Technology is humanity's preferred mode of evolution. And just as evolution has no purpose, no direction, no goal but is just the process by which whatever proves adept finds a niche, so also with technology. Technological change does not equal "progress" but it should enable us to evolve as conditions change and niches open and close. We humans now are not superior in any way to all the humans who came before us. Our lives are not better or worse than theirs, as different as they indeed may be. That will be the case as we continue to evolve—if we do continue to evolve.

To be sure, humans should try to use "reason" to guide this process, but whim, curiosity, revenge, and chance will play a bigger role than reason—as has always been the case. So all change is for the better for all who are prepared for the change, to the extent that is possible, and for those who are lucky. There is no worst case scenario to life. There is no best case scenario either. The challenge is to make the most of every situation you find yourself in, in part by practicing varieties of future possibilities beforehand so they are familiar when they appear.

All the old binaries are gone, or soon will be. The distinction between life and nonlife, between the organic and the mechanical, between animals, plants, microbes, fungi; between intelligences; between the static and dynamic; the environment and environed; you, me, us, others—going, going, gone—in spite of all the blood and tears billions spend now on defending and attacking each other's claims on identity.

In Old English, the term we now spell as "weird" was originally spelled as *wyrd*, and it did not mean strange or abnormal things. It meant destiny, fate, tragic if sometimes heroic behavior that leads to inevitable but unpredictable consequences. It is also in the sense of *wyrd* that we and our futures are weird. We act foolishly even when we know we shouldn't. We just can't help ourselves. We mean to do good, but oh so seldom do.

In *Hyposubjects: On Becoming Human*, Timothy Morton and Dominic Boyer (2021). believe they see waves of the futures in the behaviors of a category of beings they call "hyposubjects". If you have been paying attention to what I have been writing in the past one hundred pages or so, you will find them to be familiar old friends. Go look in the mirror. You might find a friendly face there.

- Hyposubjects are the native species of the Anthropocene and only just now beginning to discover what they may be and become.
- Like their hyperobjective environment, hyposubjects are also multiphasic and plural, not-yet, neither here nor there, less than the sum of their parts. They are in other words subscendent rather than transcendent. They do not pursue or pretend to absolute knowledge and language let alone power. Instead they play, they care, they adapt, they hurt, they laugh.
- Hyposubjects are necessarily feminist, antiracist, colorful, queer, ecological, transhuman and intrahuman. They do not recognize the rule of androleukoheteropetromodernity and the apex species behavior it epitomizes and reinforces. But they also hold the bliss-horror of extinction fantasies at bay because hyposubjects' befores, nows and afters are many.
- Hyposubjects are squatters and bricoleuses. They inhabit the cracks and hollows. They turn things inside out and work with scraps and remains. They unplug from carbon gridlife and hack and redistribute its stored energies for their own purposes.
- Hyposubjects make revolutions where technomodern radar can't glimpse them. They patiently ignore expert advice that they don't or can't exist. They are skeptical of efforts to summarize them, including everything we have just said (Morton and Boyer 2020:14–15).

## References

Anderson, Ruby. 2020. Space is gay, and it has an important lesson for us. https://www.thrillist.com/news/nation/space-is-gay-cosmos-model-of-inclusivity.

Astroaccess. 2020. Outer space is not just about humanity's future. https://astroaccess.org/about/.

Augusto, Juan, Vic Callaghan, Diane Cook, Achilles Kameas, and Ichiro Satoh. 2013. Intelligent environments: A manifesto. In *Human-centric computing and information sciences*, vol. 3, ed. J. Augusto et al., 12.
Bullinga, Marcella. 2004. Intelligent government: Invisible, automatic, everywhere. *The Futurist*, July-August, 32–36.
Butler, Philip. 2019. The black posthuman transformer: A secularized technorganic. *Journal of Futures Studies*, December.
Center for Faith-Based and Neighborhood Partnerships. 2022. https://www.hud.gov/offices/fbci.
Contreras, Russell. 2021. Afrofuturism envisions space in 2051. *Axios*, September 4.
Dator, James. A. 2012. *Social foundations of human space exploration*. New York: Springer Briefs in Space Development.
Dator, Jim. 2020. Designing new forms of governance. *Public Policy Review* 1, November.
Dator, Jim. 2006. Will America ever become a democracy? In *Democracy and futures*, ed. Mika Mannermaa, Jim Dator, and Paula Tiihonen. Helsinki: Parliament of Finland.
Davis, Erik. 1992. DeLanda destratified: Manuel DeLanda observed. *Mondo 2000* 8, 47.
ESA. 2016. Space 4.0. https://www.esa.int/About_Us/Ministerial_Council_2016/What_is_space_4.0.
European Wilderness Society. 2022. A river in New Zealand legally becomes a person. https://wilderness-society.org/a-river-in-new-zealand-legally-becomes-a-person/.
Evelein, Rose. 2019. Its time to rethink who's best suited for space travel. *Wired*, January 27. https://www.wired.com/story/its-time-to-rethink-whos-best-suited-for-space-travel/.
Finney, Ben, and Eric Jones, eds. 1985. *Interstellar migration and the human experience*. Berkeley: University of California Press.
Foust, Jeff. 2021a.The normalization of space tourism. *Space Review*, October 18.
Foust, Jeff. 2021b. Private human spaceflight become more regular, but not routine. *Space Review*, December 13.
French, Francis. 2020. Gay astronauts: A final frontier. *The Vintage Space*, October 14.
Garan, Ron. 2015. *The orbital perspective: An astronaut's view*. London: Metro Publishing.
Garrett-Bakelman, Francine E. et al. 2019. The NASA twins study: A multidimensional analysis of a year-long human spaceflight. *Science* 364, 144.
Gipson, Grace. 2019. Creating and imaging black futures through afrofuturism. In *#identity*, ed. Abigail De Kosnik, and Keith P. Feldman. Ann Arbor: University of Michigan Press.
Glenn, Jerome, and George Robinson. 1978. *Space trek: The endless migration*. New York: Stackpole Books.
Gohd, Chelsea. 2020. Charles Bolden, NASA's First Black administrator, speaks out on systemic racism. *Space.com*, June 17.
Goonatilake, Susantha. 1999. *Merged evolution: Long term implications of biotechnology and information technology*. Philadelphia: Gordon and Breach.
Holmes, Juwan. 2021. https://www.lgbtqnation.com/2021/12/cameron-bess-becomes-first-pansexual-person-go-space/.
Joy, Bill. 2000. Why the future doesn't need us. *Wired*, April 1.
Kolbert, Elizabeth. 2022. Testing the waters. Should the natural world have rights? *The New Yorker*, April.
Krieger, Martin H. 1973. What's wrong with plastic trees? *Science* 179, February 2.
Lindsay, Vachel. 1923. The eagle that is forgotten. In *Collected poems of Vachel Lindsay*. Whitefish, Montana: Kessinger Publishing.
Martin, Layla. 2021. The problem with space cowboys. *Space Review*, September 13.
McNally, Philip, and Sohail Inayatullah. 1988. The rights of robots: Technology, law and culture in the 21st century. *Futures* 20/2.
Morton, Timothy, and Dominic Boyer. 2021. *Hyposubjects: On becoming human*. London: Open Humanities Press.
Platt, John. 1981. Acceleration of evolution. *The Futurist*, 14–23, February.
Pontin, Jason. 2018. The genetics (and ethics) of making humans fit for Mars. *Wired*, August 7.

Schwab, Klaus. 2016. https://www.weforum.org/agenda/2016/01/the-fourth-industrial-revolution-what-it-means-and-how-to-respond/.
Scott-Heron, Gil. 1970. Whitey on the moon. *Small Talk at 125th and Lenox*. Atlantic Records.
Stone, Christopher B. 1972. Should trees have standing? Towards legal rights for natural objects. *Southern California Law Review* 45.
Stott, Nicole. 2021. *Back to Earth: What life in space taught me about our home planet—and our mission to protect it*. New York: Seal Press.
Szocik, Konrad, ed. 2020. *Human enhancements for space missions: Lunar, Martian, and future missions to the outer planets*. New York: Springer.
Tenrikyo. nd. https://www.tenrikyo.or.jp/eng/.
The Mason Lab. 2020. https://www.masonlab.net/.
Timmers, Samantha. 2020. *On the advantages of the disabled in space*. Senior Honors Theses: University of Louisville. College of Arts & Sciences.
Torres, Phil. (forthcoming). *Human extinction: A history of thinking about the end of the world*.
Weibel, Deana L. 2020. "Maybe you were put here to be the answer". Religious overtones in the new space force recruitment video. *Space Review*, May 11.
White, James. 1987. *The overview effect and human exploration*. Boston, Mass.: Houghton Mifflin.
White, James. 2019. *The cosma hypothesis: Implications of the overview effect*. Littleton, CO: Morgan Brook Media.
Yamada, Tadakazu, ed. 1988. *Cosmos-life-religion: Beyond humanism*. Tenri, Japan: Tenri University Press.

# Chapter 14
# Weirding the Queer

**Abstract** A nod of thanks to queer pioneers for queering everything in this and all possible worlds, including fungi and lichens.

**Keywords** Fungi · Lichens · Queer · Weird

At various points in this essay I have expressed a debt of gratitude to queer theory for tossing so many moldy dichotomies out of the closet; for queering everything. Nonetheless, I was quite startled when reading "Our Silent Partners", by Zoë Schlanger (2021) when I ran across this passage: "Sheldrake and Spribille are both proponents of using queer theory to find ways of understanding fungi where science has yet to draw a map. 'The human binary view has made it difficult to ask questions that aren't binary,' Spribille says. "Our strictures about sexuality make it difficult to ask questions about sexuality, and so on.... And this makes it extremely difficult to ask questions about complex symbioses like lichens.' As I read *Entangled Life*, I found myself thinking that perhaps this has held us back from seeing what is abundantly evident: that queerness, in its embrace of infinite variation and fluctuating identities, has always been natural." Among the articles Sheldrake cites is David Griffiths, "Queer Theory for Lichens" (Griffiths 2015).

"'We are all lichens,' the biologist Scott Gilbert and his colleagues wrote. Titled "A Symbiotic View of Life: We Have Never Been Individuals," the paper argues that "human bodies cannot be thought of as individual organisms. Rather, we are dynamic colonies housing a shifting community of trillions of bacteria and fungi that perform many vital functions and determine our health (including a 'microbial cloud' that overflows our body, hovering, at all times, in the air around us). From this point of view, 'I' has always truly been 'we.' While much research and discussion has been devoted to the bacterial constituents of our microbiomes in recent years, fungi play just as vital a part in governing our internal ecosystems, particularly our digestive and immune systems. Despite this, fewer than one percent of all scientific papers published on the microbiome concerns its fungal constituents" (Gilbert 2012).

The futures seem to belong to lichens, fungi, cyborgs and their descendants. Please join me as we shed our old identities and containers, and merge on towards still weirder worlds than we can possibly imagine. All together now: "When the

 
J. Dator, *Beyond Identities: Human Becomings in Weirding Worlds*,
Anticipation Science 7, https://doi.org/10.1007/978-3-031-11732-9_14

world gets weird, the weird get weirder till what's weird gets normal and what's normal weirder still."

## 14.1 Coda

OK. I admit it. I am not a robot, a computer, a dator. I am a very ordinary human becoming.

I also do know who my father is because my mother felt impelled to tell me what she believed to be the truth when she was in her late 70s. She concluded that I needed to know after I experienced a brief puzzling interaction with the family of the person I had been told all my life was my father who had drowned while I was barely a year old. While I had met this family once as a very small child, my only other encounter with them, when I was in my late 50s, was by chance, and fraught with very awkward confusion and misunderstandings as to who I was, or thought I was, in relation to them. So Mother rushed to Honolulu to straighten me out once and for all before I made matters any worse.

It turned out that my actual father was a person I knew very slightly with whom I had a brief conversation as a young adult that nonetheless resulted in one of the most important decisions I ever made; one that had a deep and lasting impact on my life and my image of myself: being a high school football coach himself, he casually suggested that I walk-on for the Stetson University football squad, and then pulled strings that made that possible. As a consequence, I played a few downs during the upset victory by Stetson over Arkansas State University in the Tangerine Bowl in Orlando, Florida. I played in a Bowl Game! I was never a great athlete—not even a particularly good one—but I was on the football squad for two years (I graduated early before I could play more) and was always very proud of that. It gave me a sense of balance and wholeness at being able to excel academically that was important to me as well as to hold my own athletically in a brutal sport that was important to others. Of course, only White people attended Stetson then and played on the team. If the squad had represented the athletic diversity available (as it does now), I would have been sitting in the stands rooting them on.

It turns out that my father was a second generation Italian-American and my mother (as I knew) from longtime English-American stock. That would seem to mean I am vanilla White—but remember that Italian-Americans were still suspect as scum and grifters when my father was young though he was firmly successfully middle class by the time he died.

I had no other interaction with my actual father or father's family, and none of them knew I existed, I believe. So I stick to my story as a fatherless child reared primarily by three independent, capable women—along with a host of other people who made me what I became. To them all I once again express my gratitude, but vigorously affirm that my loyalties nonetheless lie not in the past but firmly with the futures and human becomings.

So might yours. Why not give it a try?

## References

Gilbert, Scott. 2012. A symbiotic view of life: We have never been individuals. *Quarterly Review of Biology*.
Griffiths, David. 2015. Queer theory for lichens. *UnderCurrents*, 19.
Schlanger, Zoë. 2021. Our silent partners. *The New York Review*, October 7.

The manufacturer's authorised representative in the EU is Springer Nature Customer Service Centre GmbH, Europaplatz 3, 69115 Heidelberg, Germany. If you have any concerns regarding our products, please contact ProductSafety@springernature.com

Printed and bound by CPI Group (UK) Ltd, Croydon, CR0 4YY
28/09/2026
02224091-0001